ENVIRONMENTAL SCIENCE, ENGINEERING AND TECHNOLOGY

ENVIRONMENTAL MANAGEMENT

ECOSYSTEMS, COMPETITIVENESS AND WASTE MANAGEMENT

ENVIRONMENTAL SCIENCE, ENGINEERING AND TECHNOLOGY

Additional books and e-books in this series can be found on Nova's website under the Series tab.

ENVIRONMENTAL SCIENCE, ENGINEERING AND TECHNOLOGY

ENVIRONMENTAL MANAGEMENT

ECOSYSTEMS, COMPETITIVENESS AND WASTE MANAGEMENT

MIGUEL FISCHER
EDITOR

NOTICE TO THE READER

Library of Congress Cataloging-in-Publication Data

ISBN: 978-1-68507-019-9

Published by Nova Science Publishers, Inc. † New York

CONTENTS

PREFACE

As defined by the EPA, an environmental management system refers to a set of processes and practices that enable an organization to reduce its environmental impacts and increase its operating efficiency. Accordingly, this book presents five chapters that introduce unique perspectives relating to the concept of environmental management. Chapter one describes the results of a large-scale study on the structure and governance of diverse ecosystem services of Bulgarian farms. Similarly, Chapter two applies a holistic approach in the assessment of the competitiveness of agricultural holdings in Bulgaria as a whole, as well as in terms of their different specialization. Chapter three includes information that enables the adequate identification of mine waste deposits, in order to evaluate the impact on human and ecological health, and to implement suitable alternatives for their environmental management. Chapter four reviews the biomonitoring strategies employed to evaluate the negative environmental impact of mining waste, the remediation alternatives for mining-polluted sites, and environmental management approaches. Lastly, Chapter five presents suggestions of actions for public managers and entrepreneurs of the solid waste sector with the perspective of process automation, waste destination, sustainability of the planet, and the reduction of waste production.

Chapter 1 - The products and the variety of direct and indirect benefits that humans receive from nature and the various ecosystems (agricultural, forest, grass, mountain, river, marine, etc.) are commonly known as ecosystem services. Agricultural ecosystems of different types and their specific “agro-ecosystem” services are among the most widespread in the world. In recent years increasing attention is given to the system of (“good”) governance as a key to achieving public, collective, corporate, and private goals in relation to conservation and improvement of (agro)ecosystem services. Nevertheless, in Bulgaria, like in many other countries, there are few studies on the amount and importance of agro-ecosystem services, and the specific mechanisms, modes, factors, and efficiency of their management. This chapter tries to fill the gap and presents the results of a large-scale study on the structure and governance of diverse ecosystem services of Bulgarian farms. Firstly, it identifies the type, amount, and importance of various (provisional, economic, recreational, aesthetic, cultural, educational, supporting, water and air purification, biodiversity preservation, climate regulation, etc.) ecosystem services maintained and “produced” by the Bulgarian farms of different juridical type, size, specialization, and location. The study has found out that country’s farms provide a great number of essential ecosystem services among which provisioning food and feed, and conservation of elements of the natural environment prevail. Secondly, it identifies and assesses the efficiency and complementarities of specific modes and mechanisms of governance of ecosystem services used by the Bulgarian farms. The study had found out that a great variety of private, market, collective, public and hybrid modes of governance of farm activity related to agroecosystem services are applied. There is significant differentiation of employed managerial forms depending on the type of ecosystem services and the specialization of agricultural holdings. Furthermore, the management of agroecosystem services is associated with a considerable increase in the production and transaction costs of participating farms as well as big socio-economic and environmental effects for agricultural holdings and other parties. The factors that mostly stimulate the activity of Bulgarian agricultural producers for protection of (agro)ecosystems and their services are participation in public support programs, access to farmers' advice, professional training, available information, and innovation, received direct subsidies from EU and national government, personal conviction and satisfaction, positive experience of others, long-term and immediate benefits for the farm, and integration with suppliers, buyers, and processors. The suggested holistic and interdisciplinary framework for analyzing the system of management of agro-ecosystem services is to be further extended and improved, and more widely and periodically applied in the future. The later requires systematic in-depth multidisciplinary research in this new area, as well as the collection of original micro- and macro information on ecosystem survives, and forms, efficiency, and factors of their management. The accuracy of analyzes is to be improved by increasing representativeness through enlarging the number of surveyed farms and related agents, applying statistical methods, special "training" of participants, etc. as well as improving the official system for collecting agricultural, agro-economic, and agri-environmental information in the country.

Chapter 2 - In an effort to fill the existing gap regarding the definition of the competitiveness of agricultural holdings and the ways to measure it, the present research applies a holistic approach in the assessment of the competitiveness of agricultural holdings in Bulgaria as a whole, as well as in terms of their different specialization. Despite its importance and the continuing debates on the topic, there is still no consensus on what the competitiveness of farms is; how to measure the competitiveness of different organizations in agriculture; what the absolute and comparative competitiveness of different types of farms is; which are the critical factors for increasing the competitiveness at the current stage of development, etc. The multi-criteria assessment found that although the level of competitiveness of Bulgarian farms is overall good, more than a third of all farms in the country show a low level of competitiveness. This can largely be attributed to their low adaptive potential and economic efficiency. The most competitive farms are those specializing in the beekeeping sector, followed by field crops, mixed livestock and mixed crop production, while farms specializing in grazing livestock are the least competitive. The proposed approach should be improved and applied more widely and periodically, increasing its accuracy and representativeness. The latter requires close cooperation with producer organizations, the National Agricultural Advisory Service (NAAS) and other stakeholders, as well as improvements to the agricultural information collection system in the country.

Chapter 3 - Mining activities have more than 450 years of tradition in Mexico. This extractive activity has great relevance in the country's economy and development. However, mining activity and the processes used for the extraction of metallic minerals of economic interest generate a large amount of wastes, called mine-tailings, which contain a complex mixture of heavy metals. It has been estimated that there are many abandoned mine-tailings throughout Mexico due to the absence of legal regulations in the years before 2004, which implies a threat to surrounding ecosystems and human populations. In this chapter, a diagnosis of the current situation of the mine-tailings in the central region of Mexico is presented, 42 studies into mine tailing characterization were reviewed, published between 2005 and 2020. In these studies, a total of 27 different mine tailings were identified, possibly amounting 137 million tons of mining wastes, rich in complex mixtures of potentially toxic elements such as arsenic, cadmium, chromium, copper, iron, lead and zinc, whose bioavailability is an important concern for environmental and human health. The information presented in this chapter enables the adequate identification of these mine waste deposits, in order to evaluate the impact on human and ecological health, and to implement suitable alternatives for their environmental management.

Chapter 4 - Mining is one of the most important extractive activities worldwide. It has a significant impact on the economic development of different regions in many countries and supplies the growing demand for mineral resources for various industries. However, this economic activity is well recognized as an environmental hazard due to the release of several pollutants to nature. Among the most relevant environmental concerns related to mining activities are solid and liquid waste generation, including mine tailings and acidic drainage. Adequate management of such waste is an important task for avoiding negative impacts on the environment. Biomonitoring strategies in or near sites impacted by mining activities give information on the overall environmental health, enabling the identification of bioaccumulation and biomagnification processes, and opportunities for remediation of the mining-impacted sites and implementation of suitable waste management strategies. In the present chapter, the authors review the biomonitoring strategies employed to evaluate the negative environmental impact of mining waste, the remediation alternatives for mining-polluted sites and environmental management approaches.

Chapter 5 - Approximately one million people die each year in the world because of chemical contamination caused by solid waste disposal. From 1991 to 2000, the amount of solid waste produced in the world was 0.68 billion tons, from 2010 to 2018 it increased to 1.3 billion tons, and it is estimated that by 2025 it will be 2.2 billion tons. Solid waste includes food waste, paper, plastics, glass, textiles, metals, wood, leather, and others; and is produced in human activities, both in households and organizations, and requires handling, storage, collection, and disposal. The main challenges identified in the literature on solid waste management are difficulty in measuring and controlling environmental and social impacts, lack of legislation applied to waste and of specialized workers, structural and budgetary deficiency, and lack of management-oriented strategies and statistical tools. Within this scenario, the objective of this chapter is to propose recommendations for developing solid waste management based on the identification of scientific opportunities and challenges in the literature. Based on what was identified, benchmarking was performed, and recommendations were proposed to overcome the challenges of solid waste management. The main scientific contribution of this chapter is the deepening and expansion of the literature on solid waste management based on the solutions to the challenges, which will help and stimulate future research on the topic. As for the applied contribution, this work presents suggestions of actions for public managers and entrepreneurs of the solid waste sector with the perspective of process automation, waste destination, sustainability of the planet, and the reduction of waste production.

In: Environmental Management
Editor: Miguel Fischer
ISBN: 978-1-68507-019-9

Chapter 1

A Study on Structure and Governance of Ecosystem Services of Bulgarian Farms

Hrabrin Bachev*
Institute of Agricultural Economics, Sofia, Bulgaria

Abstract

The products and the variety of direct and indirect benefits that humans receive from nature and the various ecosystems (agricultural, forest, grass, mountain, river, marine, etc.) are commonly known as ecosystem services. Agricultural ecosystems of different types and their specific "agro-ecosystem" services are among the most widespread in the world. In recent years increasing attention is given to the system of ("good") governance as a key to achieving public, collective, corporate, and private goals in relation to conservation and improvement of (agro)ecosystem services. Nevertheless, in Bulgaria, like in many other countries, there are few studies on the amount and importance of agro-ecosystem services, and the specific mechanisms, modes, factors, and efficiency of their management.

---

* Corresponding Author's E-mail: hbachev@yahoo.com.

This chapter tries to fill the gap and presents the results of a large-scale study on the structure and governance of diverse ecosystem services of Bulgarian farms. Firstly, it identifies the type, amount, and importance of various (provisional, economic, recreational, aesthetic, cultural, educational, supporting, water and air purification, biodiversity preservation, climate regulation, etc.) ecosystem services maintained and "produced" by the Bulgarian farms of different juridical type, size, specialization, and location. The study has found out that country's farms provide a great number of essential ecosystem services among which provisioning food and feed, and conservation of elements of the natural environment prevail.

Secondly, it identifies and assesses the efficiency and complementarities of specific modes and mechanisms of governance of ecosystem services used by the Bulgarian farms. The study had found out that a great variety of private, market, collective, public and hybrid modes of governance of farm activity related to agroecosystem services are applied. There is significant differentiation of employed managerial forms depending on the type of ecosystem services and the specialization of agricultural holdings. Furthermore, the management of agroecosystem services is associated with a considerable increase in the production and transaction costs of participating farms as well as big socio-economic and environmental effects for agricultural holdings and other parties.

The factors that mostly stimulate the activity of Bulgarian agricultural producers for protection of (agro)ecosystems and their services are participation in public support programs, access to farmers' advice, professional training, available information, and innovation, received direct subsidies from EU and national government, personal conviction and satisfaction, positive experience of others, long-term and immediate benefits for the farm, and integration with suppliers, buyers, and processors.

The suggested holistic and interdisciplinary framework for analyzing the system of management of agro-ecosystem services is to be further extended and improved, and more widely and periodically applied in the future. The later requires systematic in-depth multidisciplinary research in this new area, as well as the collection of original micro- and macro information on ecosystem survives, and forms, efficiency, and factors of their management. The accuracy of analyzes is to be improved by increasing representativeness through enlarging the number of surveyed farms and related agents, applying statistical methods, special "training" of participants, etc. as well as improving the official system for collecting agricultural, agro-economic, and agri-environmental information in the country.

Keywords: ecosystems, services, governance, efficiency, agriculture, farms, Bulgaria

INTRODUCTION

Ecosystem services are widely known as products and other benefits that humans receive from natural ecosystems (MEA, 2005). The agricultural ecosystems and their specific "agro-ecosystem" services are widespread in Bulgaria and internationally (EEA; FAO). Since the introduction of this concept in the last years of the 20th century, (agro) ecosystem services have been intensively promoted, studied, mapped, evaluated, and managed (Adhikari et al.; Allen et al.; Boelee; De Groot et al.; EEA; FAO; Fremier et al.; INRA; Gao et al.; Garbach et al.; Gemmill-Herren; Habib et al.; Kanianska; Lescourret et al.; Laurans and Mermet; Marta-Pedroso et al.; MEA; Munang et al.; Nunes et al.; Novikova et al.; Petteri et al.; Power; Scholes et al.; Tsiafouli et al.; Van Oudenhoven; Wang et al.; Wood et.al.; Zhan).

Despite growing environmental issues, and increasing public and private interests, the scientific studies in that new area are still a "work in progress." Research is commonly limited to a certain type of agro-ecosystem services (e.g., plant pollination, biodiversity conservation, etc.), a particular ecosystem (e.g., Zapadna Stara Planina, etc.), a single aspect of the management (agronomic, technological, etc.), a specific form of governance (a public support scheme, organic agriculture, etc.), a separate level of management (farming organization, region, etc.), the specific type of costs and benefits (production, direct, etc.), etc. At the same time, the importance of effective management ("good" governance) for conservation and sustainable provision of ecosystem services in general and of a certain type has been broadly recognized by the academic community, policymakers, interest groups, professional and business organizations, and the public at large (Bachev; EEA; FAO; UN).

In Bulgaria, research on economic and other issues related to agroecosystem services are at the beginning stage and mostly at "conceptual and methodological" level (Bachev; EEA; Grigorova and Kazakova; Kazakova; Nedkov; Nikolov; Todorova; Tchipev et al.; Yordanov et al.). Besides, there very few studies on dominating modes of governance at the current stage of development and fundamental transformation of EU CAP (Bachev; Bachev et al.; Todorova). This article fills the gap and presents the result of the first-in-kind "large-scale" study on structure and governance of ecosystem services of Bulgarian farms at current stage of development.

1. Methods and Data

"Agrarian" ecosystems and "agrarian" ecosystem services are those associated with the agricultural "production" (Bachev, 2020). The hierarchical system of agroecosystems includes multiple levels (from individual farm plot/section, area, micro-region, macro-region, etc.) (Figure 1) while their (ecosystem) services are classified into different categories (provisional, economic, recreational, aesthetic, cultural, educational, supporting, biodiversity conservation, water purification and retention, flood and fire protection, climate regulation, etc.) (MEA).

The multiple levels hierarchy of agro-ecosystems and their services in Bulgaria is presented in Figure 1. Individual farm is the main organizational unit in agriculture that manages resources, technologies and activities and produces a variety of products, including the positive and negative services of agro-ecosystems (Bachev, 2009, 1010; 2020). The governance of agro-ecosystem services is an integral part of the management of agricultural farm, and the farm - the first (lowest) level for agro-ecosystem services management[1].

[1] Farm borders rarely coincide with the (agro) ecosystem boundaries (Bachev).

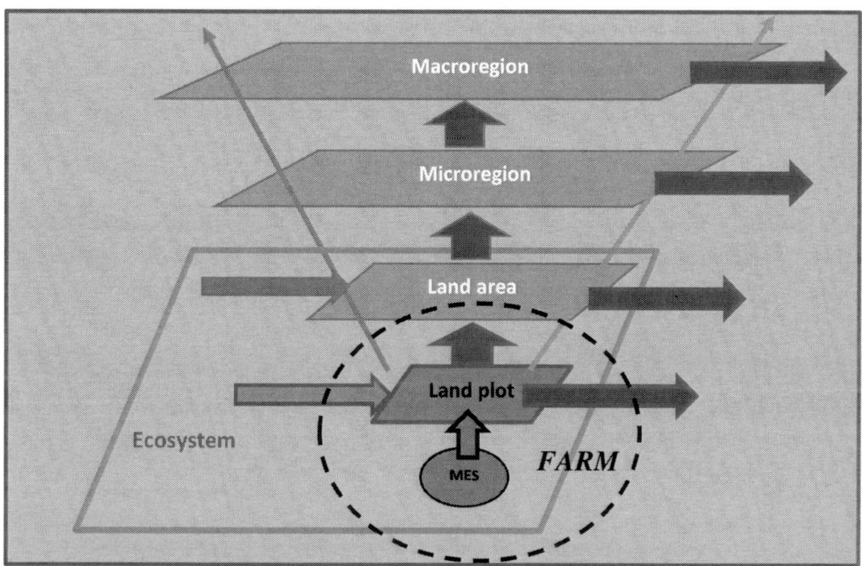

Blue – agro-ecosystem, Red – Agroecosystem Services, MES – Micro ecosystem located in the land plot, Green – Services of non-agrarian ecosystems, Dash area – Borders (activity) of individual farm.
Source: Author.

Figure 1. Hierarchy of Agro-ecosystems in Bulgaria.

The term "management of (agro)ecosystem services" refers to the management of human actions and behavior related to preservation, improving and recovery of ecosystems and ecosystem services (Bachev, 2009). The system of governance of agro-ecosystem services always includes the farm as a key element (first level) of management of agro-ecosystems and their services (Figure 1). Other agrarian and not agrarian agents (resource owners, inputs suppliers, wholesale buyers and processors, interests groups, policymakers, local and national authorities, residence and visitors of rural areas, final consumers, international organizations, etc.) also take part in the management of agroecosystem services at farms, regional, sectoral, national and international levels (Bachev, 2020).

Farmers use diverse mechanisms and modes to manage their activity and relations with other agents (Bachev, 2010; Williamson, 1996):

- internal - direct production management, own conviction of farm manager/owner, building reputation, etc.;
- market - free-market price movements, competition, etc.;
- contract - special or interlinked contracts, etc.;
- collective - cooperation, joint initiatives, etc.]
- public - public eco-contract, cross-compliance against EU subsidization, etc.

Detailed presentation of the New Institutional Economics framework for studying and evaluating generic modes of governance, and comparative advantages and disadvantages of individual forms used for ecosystem services management in Bulgarian agriculture is done in other publications (Bachev, 2009, 2011, 2012, 2020).

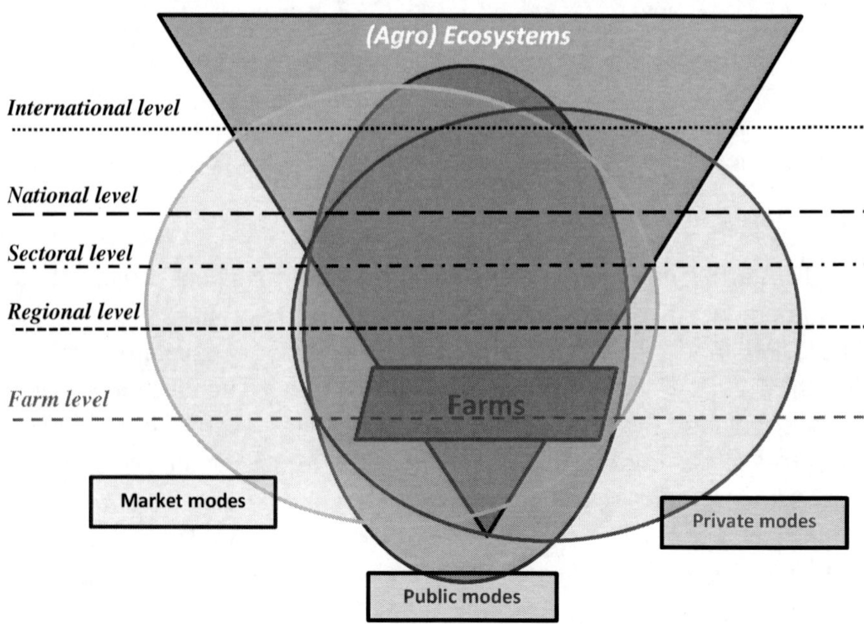

Source: Author.

Figure 2. Levels and Modes of Governance of Agroecosystem Services.

In Bulgaria there is no available statistical and other data on services provided by different type of agroecosystems. Since the individual farm is the basic unit of management of agrarian activities and provision of agro-ecosystem services, our study has focused on the (individual) *farm level* of maintenance and supply of ecosystem services. The agroecosystem services at a higher lever are evaluated as sum of agroecosystem services provided by the farms associated with the relevant (agro)ecosystems. Consequently, there is an unavoidable error from double accounting and/or uncalculated trade, synergies, complementarities and controversies of analyzed agroecosystem services of different type.

This study aims to identify modes, efficiency and factors of agroecosystem services management at the farm level in Bulgaria. In the country, there are statistical and other data for the type of agroecosystem service provided by farms and the specific forms of management applied by agricultural holdings. Therefore, a literature review and widespread practices examination has been made to prepare the list of diverse types of agro-ecosystem services maintained or provided as well as major forms of management used by the farms. A survey with the managers of 324 "typical" farms[2] of different legal type, size, production specialization, and ecological and geographical location was conducted in October 2020 to identify the structure of ecosystem services "produced" and governing modes employed. The questionnaire also gives an option to respondents to add specific services provided and managerial forms practiced.

Surveyed farms account for almost 0,5% of all registered agricultural producers in the country. The structure of studied holdings approximately correspond to the real structure of farms in Bulgaria. The classification of agricultural holdings has been done according to official classification in the country and EU. The subsectors, regional, national, etc. summaries are arithmetic averages of data provided by the individual farms belonging to respective agro-systems.

The assessments of the farm manages about type, amount, and importance of agroecosystem services they maintain or produce give good insights on the state and efficiency of agroecosystem services in the country. The asymmetry of information is quite big in the area and farmers are among the most informed actors about agricultural efforts

---

[2] The author is grateful to all managers of the surveyed farms for the information provided, and to the NAAS and the cooperating producers' organizations for the assistance.

and contribution toward (agro)ecosystem services. However, the managers estimates also reflects the "personal" (subjective) knowledge and perceptions of the farmers on agroecosystem services, and their values, the efforts rather than output and impacts, etc. The objectivity of the study would enhanced during the next stage of the study when farms representations will be increased and their assessments complemented ("corrected") with estimates of stakeholders, consumers, experts, etc.

2. Type and Amount of Farms' Ecosystem Services

The conducted survey allowed to make a detailed map of the agro-ecosystem services of different types provided by agricultural producers, as well as to determine the structure and volume of the services of the agro-ecosystems of various types. The share of farms involved in activities related to the provision of agro-ecosystem service of a certain kind gives a good idea of the volume of "produced" service of that type.

The majority of Bulgarian farms participate in the "Production of products (fruits, vegetables, flowers, etc.) for direct human consumption" (59.3%), which is one of the main "services" of agro-ecosystems in the country (Figure 3). A significant part of the farms also "Produce raw materials (fruits, milk, etc.) for the food industry" (15.4%). Other "production" services in which a smaller part of the farms participate are "Production of animal feed" (8.6%), "Own processing of agricultural products" (6.17%), "Production of seeds, saplings, animals, etc. for farms" (4.3%) and "Production of raw materials for cosmetic, textile, energy, etc. industry" (3.09%).

Other "production" services of agroecosystems, in which a relatively small part of agricultural producers participate, are "Provision of services to other farms and agricultural organizations" (2.47%), "Provision of services to end users (riding, fruit picking, etc.)" (1.85%), "Provision of tourist and restaurant services" (0.62%) and "Production of bio, wind, solar, etc. energy" (0.62%).

Other important services of the agro-ecosystems, in which "supply" a large part of the agricultural holdings participate, are "Hiring workers"

(11.11%) and “Providing free access on the farm to outsiders” (10.49%).

Relatively many of the farms are also involved in the protection and preservation of technological, biological, cultural and other heritage - "Preservation of traditional crops and plant varieties" (6.17%), "Preservation of traditional species and breeds of animals" (7.41%), "Preservation of traditional methods, technologies and crafts" (6.17%), "Preservation of traditional products" (6.17%), "Preservation of traditional services" (5.55%), "Preservation of traditions and customs" (3.7%) and "Preservation of historical heritage" (1.23%).

A major part of agro-ecosystem services consists in preserving, restoring and improving the elements of the natural environment - soil, water, air, gene pool, landscape, plants and animals, etc. The activity of a large part of the agricultural holdings is aimed at the production of this type of agro-ecosystem services - “Disease control (measures)” (24.69%), “Pest control (measures)” (19.75%), “Protection of natural biodiversity" (18.52%), "Protection and improvement of soil fertility" (16.67%), "Protection from soil erosion" (13.58%), "Protection and improvement of soil purity" (12.34%), "Protection of surface water” (11.73%),“ Protection of groundwater purity” (9.88%),“ Ffire protection (measures)” (8.64%), and “Protection of plant and/or animal gene pool” (8.02%).

A relatively smaller part of the farms are also included in “(Measures for) water conservation and saving” (5.55%), “(Measures for) regulation of the correct outflow of water” (4.32%), "Preservation of air quality" (4.32%), "Preservation of traditional scenery and landscape" (3.7%), "Improvement (aesthetics, aroma, land use, etc.) of scenery and landscape "(3.09%), "(Measures for) regulation and improvement of the microclimate" (3.09%), "Flood protection (measures)" (2.47%), and “Greenhouse gas emission reduction (measures)” (2.47%), and "(Measures) for storm protection” (1.85%).

One of the essential services of agroecosystems is the recovery and recycling of "waste" from various activities in the sector and other industries. The main activity of many farms in this regard is "Use of manure on the farm" (13.58%), and to a lesser extent "Reuse and recycling of waste, composting, etc." (3.09%) and "Use of sludge from water treatment on-farm” (0.62%).

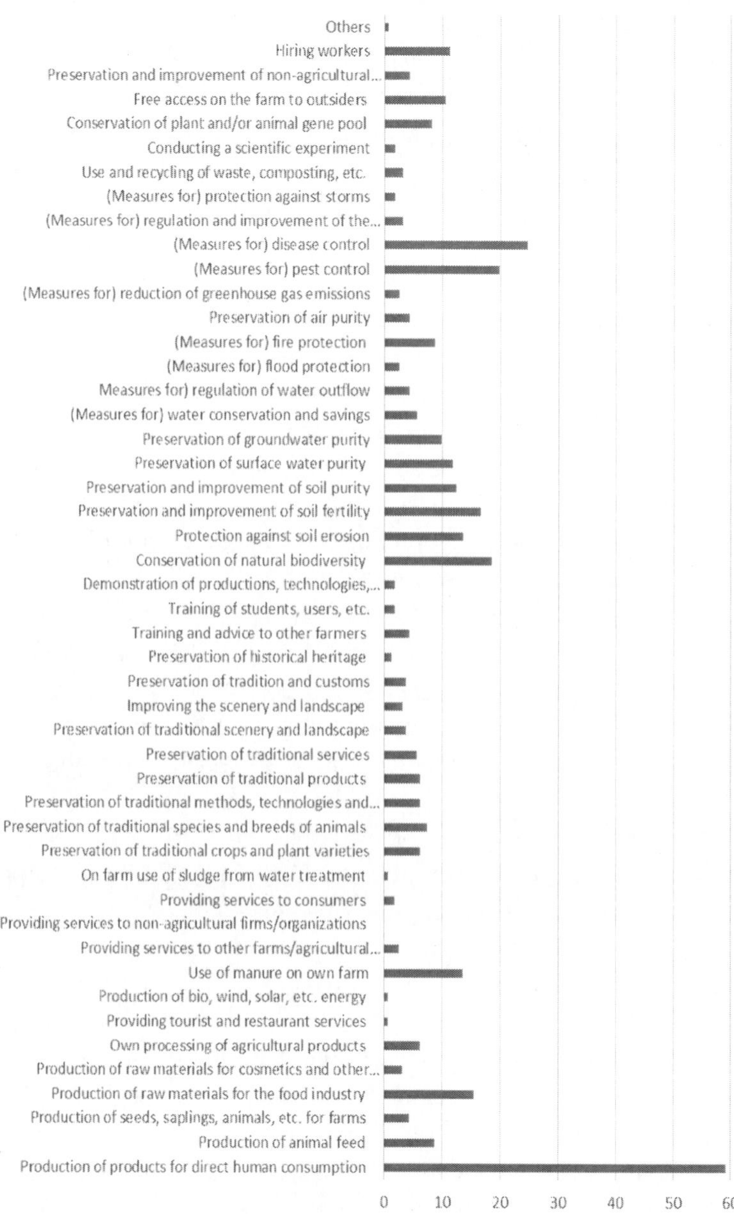

Source: Survey of agricultural producers, 2020.

Figure 3. Share of farms participating in (supporting) the preservation or production of different types of agro-ecosystem services in Bulgaria (percentages).

Agri-ecosystems also make a significant contribution to training farmers and non-agricultural agents, conducting scientific experiments, demonstrating innovation, and so on. In such educational, scientific and innovative services participate a smaller part of the agricultural producers - "Training and advice of other farmers" (4.32%), "Training of students, consumers, etc." (1.85%), "Demonstration of production, technologies, innovations, etc." (1.85%) and "Conducting a scientific experiment" (1.85%).

Agroecosystems also contribute to the "Protection and improvement of non-agricultural (forest, lake, urban, etc.) ecosystems" with 4.32% of farms in the country engaged in such efforts.

The extent of participation of supplying farms in the preservation or production of agro-ecosystem services is not equal. For most agri-ecosystem services, the holdings involved in the activities do so "To a large extent' (Figure 4). Therefore, "permanent" investments in agri-ecosystem services and "specialization" in the provision of agro-ecosystem services of a certain type to participating farms can be considered.

In some agro-ecosystem services, the share of farms involved to a large and small extent is equal - for example in the use of manure on the farm, the provision of services to other farms and agricultural organizations, (flood protection) measures, and the hiring of workers. Therefore, a significant proportion of farms are either in the process of initially "entering" (testing, studying, adapting, etc.) in the related agro-ecosystem services, or participate in this supply as ancillary or related to the main activity.

With regard to three main types of agro-subsistence services, most of the farms involved in their supply do so to a small extent – on farm using sludge from water treatment, training of students, consumers, etc., and use and recycling of waste, composting, etc. This is a sign of either the initial entry into or exit from this activity, or the inefficiency of its further expansion (intensification) by practicing farms.

The unequal participation of farmers in the provision of agro-ecosystem services of different types and unlike degrees of involvement in such activities shows the need to take measures to improve, diversify and intensify this activity through training, information, exchange of experience, public incentives, etc.

Source: Survey of agricultural producers, 2020.

Figure 4. Extent of participation (support) of farms in preservation or production of various types of agro-ecosystem services in Bulgaria.

There are significant differences and deviations from the average level in the participation of agricultural holdings in the preservation and supply of agro-ecosystem services in the main geographical and agricultural regions of the country (Figure 5).

North-western region surpasses the other regions in terms of share of farms contributing to agro-ecosystem services for production of raw materials for the food industry (17.5%), own processing of agricultural products (12.5%), provision of tourist and restaurant services (2.5%), provision of services to end-users (5%), and protection and improvement of soil fertility (22.5%).

The North Central region is a champion in terms of farm participation in the preservation of traditional crops and plant varieties (16.67%), preservation of traditional methods, technologies and crafts (10%), preservation of traditional products (10%), (measures for) fire protection (13.33%) and protection of plant and /or animal gene pool (13.33%).

The Northeast region is the largest supplier of the following agroecosystem services - production of animal feed (15.79%), production of seeds, saplings, animals, etc. for farms (10.53%), production of raw materials for cosmetics, etc. industries (15.79%), production of bio, wind, solar, etc. energy (5.26%), (measures for) pest control (42.1%), (measures for) disease control (47.37%), conducting a scientific experiment (5.26%), providing free access on the farm to outsiders (15.79%) and hiring workers (21.05%).

Southwestern region has a leading position only in terms of three agroecosystem services - production of animal feed (13.33%), provision of services to other farms and agricultural organizations (6.67%) and conservation of traditional species and breeds of animals (13.33%).

South Central region is the largest producer of many agro-ecosystem services - production of products for direct use by human (82.35%), use of manure on the farm (23.53%), preservation of traditional species and breeds of animals (14.7%), preservation of traditional methods, technologies and crafts (11.76%), preservation of traditional services (14.7%), preservation of traditional scenery and landscape (11.76%), improvement of scenery and landscape (8.82%), preservation of tradition and customs (8.82%), training and advice of other farmers (11.76%), training of students, consumers, etc. (8.82%), demonstration of productions, technologies, innovations, etc. (2.94%), protection of natural biodiversity (26.47%), protection against soil erosion (29.41%), protection and improvement of soil fertility (26.47%), protection and improvement of soil purity (20.59%), protection of purity of surface waters (20.59%), protection of groundwater purity 17.65%, (measures for) conservation and savings of water (14.7%), protection of air purity (11.76%), (measures for) reduction of greenhouse gas emissions (8.82%), (measures for) pest control (23.53%), (measures for) control of diseases (35.29%), (measures for) regulation and improvement of the microclimate (11.76%), (measures for) protection against storms (8.82%), use and recycling of waste, composting, etc. (14.7%), conducting a scientific experiment (5.88%), protection of plant and /or animal gene pool (11.76%), protection and improvement of non-agricultural ecosystems (8.82%) and employment of workers (20.59%).

Southeast region is a leader in terms of production of products for direct human consumption (66.67%), protection of natural biodiversity (29.17%), protection against soil erosion (25%), (measures to) regulate the proper outflow of water (8.33%) and fire protection (measures) (12.5%).

The large specific ecosystems in the country also differ significantly in the structure of the dominant agro-ecosystem services and in the share of the farms involved in their preservation and provision (Figure 6).

For example, the agro-ecosystem Western Stara Planina is a leader in the share of farms engaged in agro-ecosystem services related to the production of animal feed (11.54%), own processing of agricultural products (15.38%), provision of services to other farms and agricultural organizations (3.85%) and provision of services to end users (7.69%).

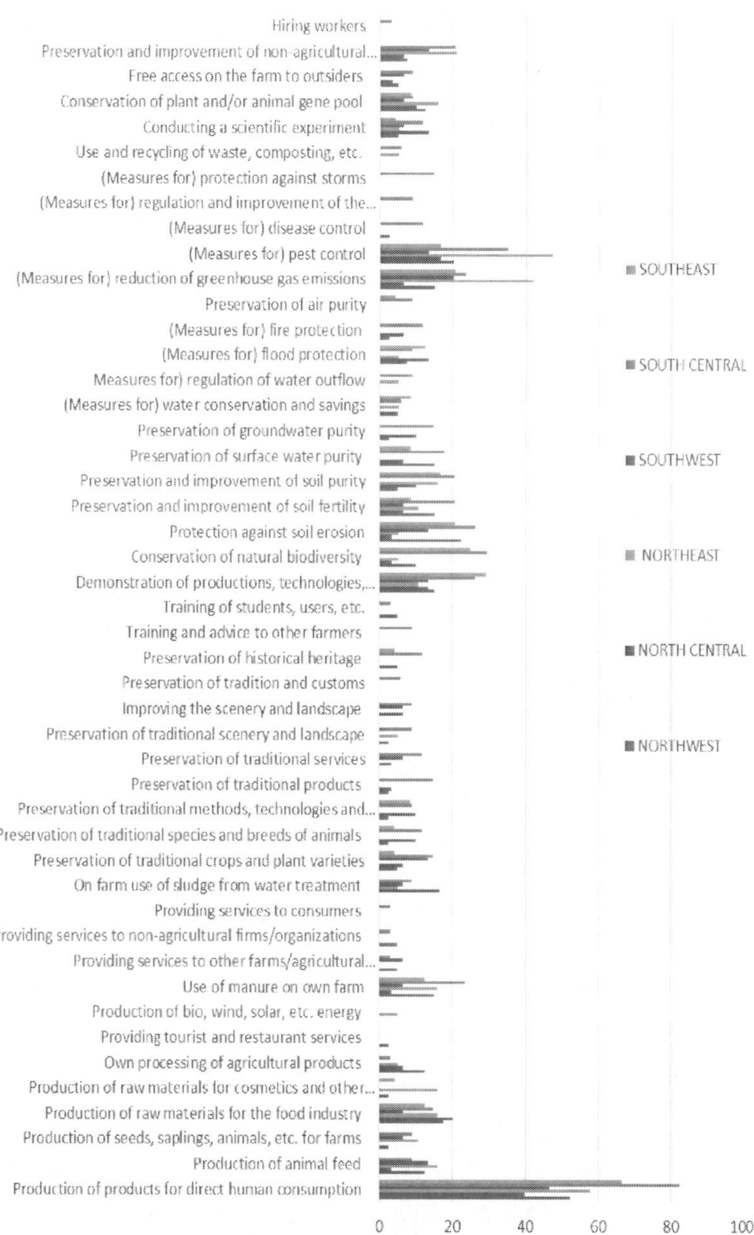

Source: Survey of agricultural producers, 2020.

Figure 5. Share of farms involved (supporting) the preservation or production of various types of agro-ecosystem services in different regions of Bulgaria (percentages).

Another studied mountainous agro-ecosystem the Rhodope Mountains is leading in the share of agricultural producers involved in the production of products for direct human consumption (78.95%), production of raw materials for the food industry (21.05%), use of manure on the farm (26.32%), preservation of traditional species and breeds of animals (10.53%), preservation of traditional methods, technologies and crafts (10.53%), preservation of traditional services (21.05%), preservation of traditional scenery and landscape (10.53%), improvement of scenery and landscape (5.26%), preservation of historical heritage (5.26%), education of students, consumers, etc. (5.26%), protection of natural biodiversity (26.32%), protection from soil erosion (31.58%), protection and improvement of soil fertility (26.32%), protection of air purity (10.53%), (measures of) reduction of greenhouse gas emissions (5.26%), (measures for) regulation and improvement of the microclimate (15.79%), use and recycling of waste, composting, etc. (10.53%), protection of plant and /or animal gene pool (15.79%), and protection and improvement of non-agricultural ecosystems (5.26%).

Agri-ecosystem Danube Plain occupies leading positions in terms of the share of farms involved in the production of raw materials for the food industry (26.92%), provision of services to other farms and agricultural organizations (3.85%), preservation of traditional crops and plant varieties (7.69%), preservation of traditional species and breeds of animals (11.54%), preservation of traditional methods, technologies and crafts (11.54%), preservation of traditional products (11.54%), preservation of traditions and customs (7.69%), demonstration of productions, technologies, innovations, etc. (3.85%), protection and improvement of soil purity (19.23%), protection of groundwater purity (23.08%), (measures for) storage and saving of water (15.38%), (measures for) fire protection (15.38%), protection of plant and /or animal gene pool (15.38%), free access on the farm to outsiders (19.23%) and hiring of workers (11.54%).

The agro-ecosystem of Dobrudja surpasses the others in terms of production of seeds, saplings, animals, etc. for farms (5.55%), production of raw materials for cosmetics and other industries (5.55%), flood protection (measures) (5.55%), fire protection (measures) (16.67%), pests control (measures) (50%), (measures for) disease control (55.56%), conducting a scientific experiment (5.56%), free access on the farm to outsiders (16.67%) and protection and improvement of non-agricultural ecosystems (5.56%).

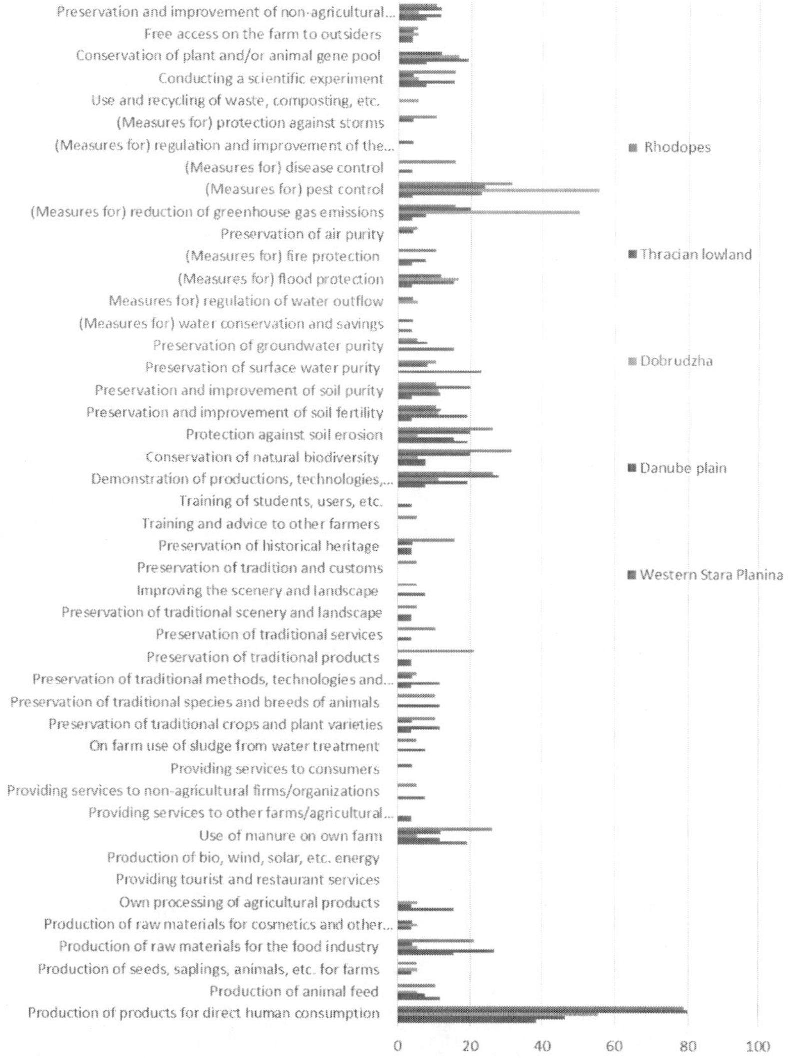

Source: Survey of agricultural producers, 2020.

Figure 6. Share of farms participating (supporting) the preservation or production of various types of agro-ecosystem services in specific ecosystems of Bulgaria (percentages).

The Thracian Lowland agroecosystem is at the forefront in terms of the share of participating farms in the production of products for direct human consumption (80%), on-farm use of sludge from water treatment (4%), conservation of natural biodiversity (28%), conservation of surface water purity (20%), storm protection (measures) (4%) and employment of workers (12%).

Farmers in the principle ecosystems of the country are also involved to varying degrees in the preservation and production of agro-ecosystem services (Figure 7). Agroecosystems in a predominantly plain region of the country are leading in the number of participating farmers in terms of production of products for direct human consumption (63.38%), provision of services to other farms/agricultural organizations (4.22%), protection from soil erosion (15.49%), protection and improvement of soil fertility (18.31%), (measures for) pest control (26.76%) and (measures for) disease control (30.98%).

Agroecosystems in the plain-mountainous regions of the country outperform the rest in terms of the share of farmers involved in the production of raw materials for cosmetics and other industries (11.43%), preservation of traditional crops and plant varieties (11.43%), preservation of traditional methods, technologies and crafts (11.43%), protection of natural biodiversity (22.86%), pest control (measures) (25.71%) and employment of workers (17.14%).

Agroecosystems in mostly mountainous regions of the country are in the best comparative position in terms of the inclusion of farms for preservation of traditional methods, technologies and crafts (11.54%), preservation of traditional services (15.38%), preservation of tradition and customs (7.69%), preservation of historical heritage (3.85%), education of students, consumers, etc. (7.69%), demonstration of productions, technologies, innovations, etc. (7.69%), (measures for) conservation and savings of water (7.69%), (measures for) regulation and improvement of the microclimate (11.54%) and hiring of workers (15.38%).

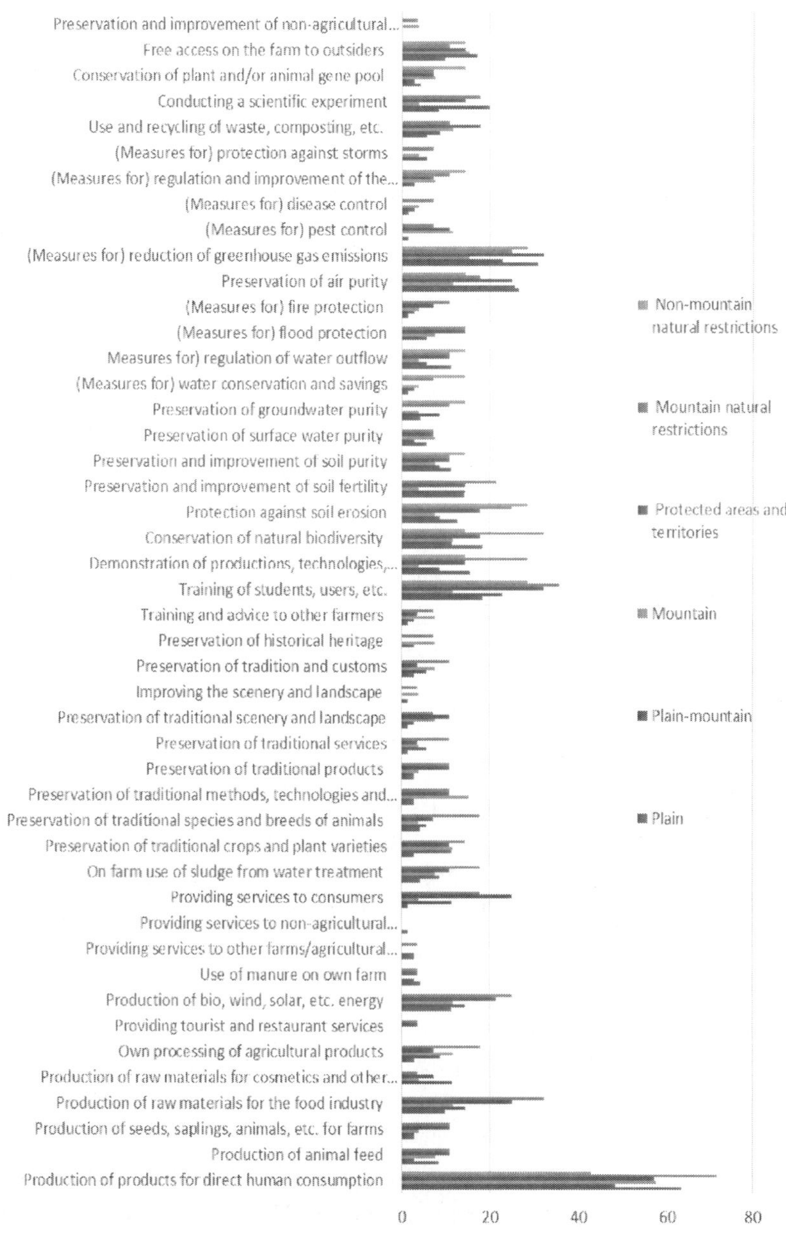

Source: Survey of agricultural producers, 2020.

Figure 7. Share of farms participating (supporting) the preservation or production of various types of agro-ecosystem services in the principle agro-ecosystems of Bulgaria (percentages).

The share of farms in agro-ecosystems in Protected areas and territories is superior to other types of agro-ecosystems in terms of production of animal feed (10.71%), production of seeds, saplings, animals and others. for farms (10.71%), production of raw materials for the food industry (25%), provision of tourist and restaurant services (3.57%), use of manure on the farm (21.43%), preservation of traditional crops and plant varieties (25%), conservation of traditional species and breeds of animals (10.71%), conservation of traditional scenery and landscape (10.71%), conservation of natural biodiversity (32.14%), conservation of air purity (14.29%), (measures for) regulation and improvement of the microclimate (10.71%) and protection of plant and/or animal gene pool (17.86%).

The agro-ecosystems in mountainous regions with natural constraints occupy leading positions in the country in terms of the share of the participating farms in the production of many agro-ecosystem services - production of products for direct human consumption (71.43%), production of animal feed (10.71%), seed production, saplings, animals, etc. for farms (10.71%), production of raw materials for the food industry (32.14%), own processing of agricultural products (17.86%), provision of tourist and restaurant services (3.57%), use of manure on the farm (25%), provision of services to end users (3.57%), preservation of traditional crops and plant varieties (17.86%), preservation of traditional species and breeds of animals (17.86%), preservation of traditional methods, technologies and crafts (14.28%), preservation of traditional products (17.86%), preservation of traditional scenery and landscape (10.71%), improvement of scenery and landscape (10.71%), preservation of tradition and customs (7.14%), training and advice of other farmers (10.71%), demonstration of production, technology, innovation, etc. (7.14%), protection of natural biodiversity (35.71%), protection against soil erosion (28.57%), protection and improvement of soil fertility (32.14%), protection and improvement of soil purity (25%), protection of purity of surface waters (21.43%), (measures for) regulation of outflow of water (10.71%), protection of air purity (14.28%), (measures for) reduction of greenhouse gas emissions (10.71%), (measures for) protection from storms (7.14%), conducting a scientific experiment (7.14%), and providing free access on the farm to outsiders (17.85%).

On the other hand, farmers in ecosystems in non-mountainous regions with natural constraints participate in the conservation and supply of a limited range of agro-ecosystem services, outperforming other agro-ecosystems in some important areas such as conservation of natural biodiversity (28.57%), protection and improvement of soil purity (28.57%), protection of the purity of the groundwater (14.28%), (measures for) regulation of the proper outflow of water (14.28%), (measures for) protection against floods (14.28%), (measures for) protection against fires (14.28%), use and recycling of waste, composting, etc. (14.28%) and protection and improvement of non-agricultural ecosystems (14.28%).

Significant differences in the preservation and provision of services of different types in the main specific and principled ecosystems of the country, and in different geographical and agricultural areas is a sign of different potential and "specialization" in supplying the main types of services from different agro-ecosystems in the country as well as of the uneven development of this activity among the agricultural producers in the different regions and ecosystems of the country.

The share of farms with different production specialization involved in the preservation and supply of agro-ecosystem services gives a good idea of the contribution of different types of production and specific agro-ecosystems to agro-ecosystem services of different types (Figure 8). For example, agro-ecosystems with field crops contribute to a relatively smaller number of agro-system services compared to other production systems in the country. However, this specific type of agro-ecosystem is superior to the others in two respects - in terms of the share of farms involved in the production of animal feed (21.43%) and fire protection (measures) (21.43%).

The vegetables and mushrooms sector is leading in the country in terms of the share of participating farms in the production of products for direct human consumption (83.33%), on-farm use of sludge from water treatment (5.55%), (measures of) storage and savings of water (11.11%), pest control (measures) (38.89%) and disease control (measures) (44.44%).

The perennials sector provides a wide variety of agro-ecosystem services, but surpasses the others only in the share of farms participating in the provision of tourist and restaurant services (1.75%) and protection against soil erosion (21.05%).

The grazing animals sector occupies leading positions in the country in terms of the share of farmers contributing to a number of agro-ecosystem services - production of raw materials for the food industry (45.45%), own processing of agricultural products (18.18%), use of manure on the farm %), provision of services to end users (9.09%), conservation of traditional species and breeds of animals (27.27%), conservation of traditional services (27.27%), protection of surface water purity (27.27%), protection of purity of air (18.18%), (measures for) reduction of greenhouse gas emissions (9.09%), use and recycling of waste, composting, etc. (18.18%), protection of plant and/or animal gene pool (27.27%), granting free access to the territory of the farm to outsiders (18.18%) and protection and improvement of non-agricultural ecosystems (27.27%).

The specialized holdings in pigs, poultry and rabbits contribute to a very limited number of agro-ecosystem services, but in several respects occupy leading positions in the country where every third producer is involved in the protection and improvement of soil purity, protection of groundwater purity, (measures for) regulating the proper flow of water, and hiring workers.

The field crops sector surpasses the others only in terms of preservation of traditional crops and plant varieties (9.09%), while those specialized in mixed livestock for two types of agroecosystem services - providing services to other farms and agricultural organizations (7.69%) and regulation and improvement of the microclimate (15.38%).

Specialized in mix crop and livestock farms participate in the supply of a wide range of agro-ecosystem services, as a relative number of participants occupy a leading position in the production of seeds, saplings, animals, etc. for farms (14.81%), preservation of traditional scenery and landscape (14.81%), improvement of scenery and landscape (11.11%), preservation of historical heritage (7.41%), training and advice of other farmers (14.81%), protection and improvement of soil fertility (25.92%), (measures for) storage and saving of water (11.11%), (measures for) protection against storms (7.41%) and conducting a scientific experiment (7.41%).

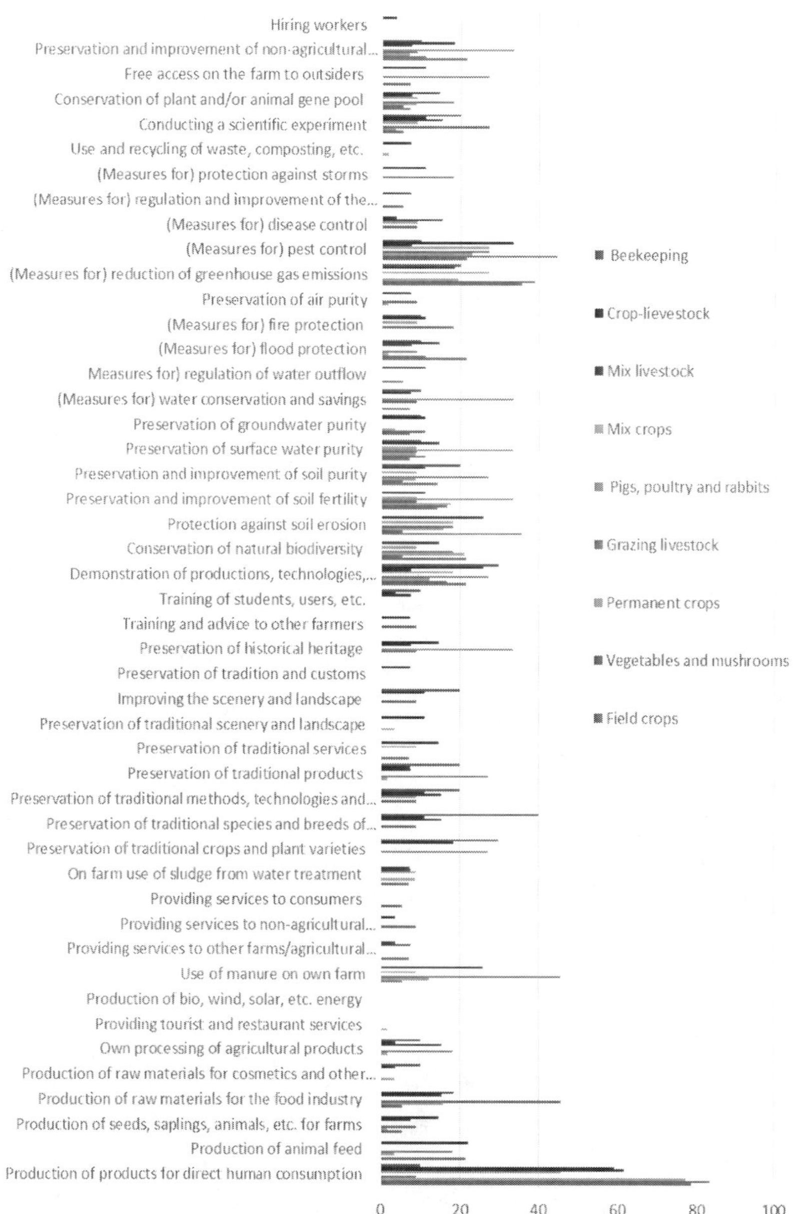

Source: Survey of agricultural producers, 2020.

Figure 8. Share of farms with different specialization participating (supporting) the preservation or production of different types of agro-ecosystem services in Bulgaria (%).

Farms specializing in bee families are characterized by the highest share of participants in the production of raw materials for cosmetics and other industries (10%), preservation of traditional species and breeds of animals (30%), preservation of traditional methods, technologies and crafts (40%), preservation of traditional products 20%, preservation of tradition and customs (20%), demonstration of productions, technologies, innovations, etc. (10%) and conservation of natural biodiversity (30%).

Significant sectoral differences in the preservation and supply of services of different types are a sign of both the different "specialization" in the supply of the main types of services from farms with different specializations and the uneven development of this activity. The later requires further research into the links between specialization and agri-ecosystem services, as well as measures to expand and diversify this activity across all farm groups.

3. Dominating Mechanisms of Management of Farms' Ecosystem Services

The survey found that a large proportion of Bulgarian farms use some specific mechanisms in making decisions about managing their activities related to agroecosystem services (Figure 9). However, a different proportion of farms apply specific mechanisms to manage the various aspects of the activity related to the provision of agro-ecosystem services. In the Production of products for direct consumption, all farms use some "special" forms[3]. A relatively large part of the farms also uses specific mechanisms in the management of Soil Protection (31.48%), Water Protection (33.95%), Biodiversity Protection (32.72%) and Landscape and Scenery Protection (20.37%). Fewer farms use specific forms to manage the supply of the other main types of agro-ecosystem services.

[3] The modes and efficiency of governance of this type of activity of Bulgarian farms have been widely studied and presented in academic literature (Bachev, 2010, 2018).

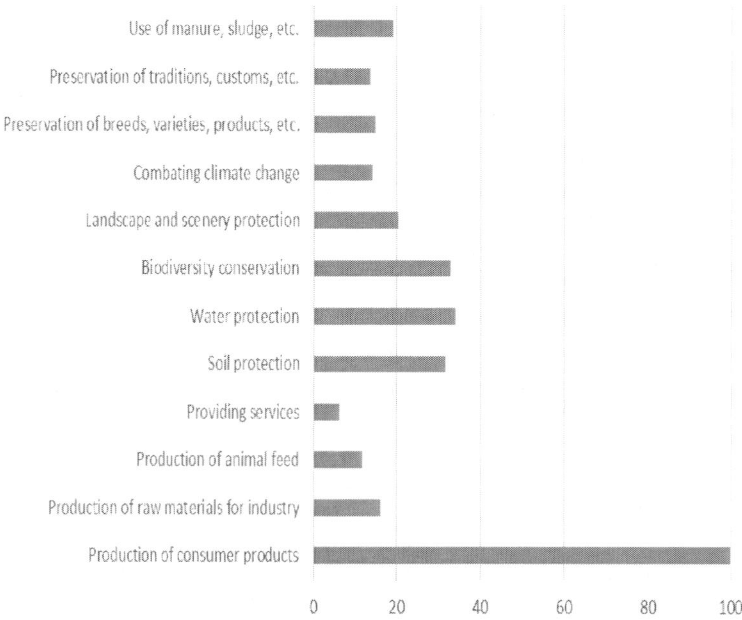

Source: Survey of agricultural producers, 2020.

Figure 9. Share of farms using specific mechanisms for decision-making of activity associated with agroecosystem services in Bulgaria (percentages).

The specific forms and mechanisms applied for the effective governance of different types of agro-ecosystem services are quite different. For most farms, independent internal (Independently by the farm) management is essential for the supply of all major agroecosystem services (Figure 10). This form is practiced by the vast majority of farms, in agro-ecosystem services with the character of "local or public goods" (inability to sell and protect rights, high specificity and uncertainty, low frequency of exchange with a particular user, etc.) - Soil protection (90.2%), Water protection (80%), Biodiversity protection (81.13%), Landscape and scenery protection (81.82%), Climate change control (78.26%), Preservation of breeds, varieties, products, etc. (87.5%) and Use of manure, sludge, etc. (90.32%). This form is least used in making management decisions concerning the production of raw materials for industry (42.31%), where there is a high dependency (specificity of the product, capacity, delivery time, location, etc.) of the particular buyer(s) and market(s) and there is a need to use more effective forms of coordination and governance.

Collective decision-making with other farmers and agents is a form that is applied by a significant part of the farms in relation to the Preservation of traditions, customs, etc. (31.82%) and a large part of them in the Production of raw materials for industry (15.38%), Water protection (14.5%), Biodiversity protection (13.21%), Landscape and Scenery protection (12.12%) and Combating climate change (13.04%). The collective form for most of these services (with the character of "local or public goods") is determined by the need for coordinated "collective action" (high dependence of assets and actions) to achieve a certain positive result. The collective organization in the production of raw materials for the industry is most often required by the need for a certain minimum volume and standardization for efficient market or vertically integrated trade (achieving efficiency in wholesale trade, compliance with the requirements of processors for quality, volume and frequency of supplies, etc.) or to oppose an existing (quasi)monopoly, etc.

Market mechanism and market prices and demand are exclusively and widely applied only to traditional (commercial) farming products and services - mostly in the Production of raw materials for industry (34.62%), Production of products for direct consumption (16.77%), and in less extent in Production of animal feed (10.53%) and Provision of services (10%). As mass and standard products are traded, the market works well and there is no need to use a more expensive special form to govern the relationship between supplier and buyer.

A special private form - Contract with a private agent/s is used when it is necessary to regulate in detail the relations of the parties due to high unilateral or bilateral dependency of assets, high frequency of transactions between the same agents, and uncertainty and risk of market trading (specification of the product, delivery time, a form of payment, interlinked transactions, a guarantee of trade between the parties, etc.). The contractual form is applied by every tenth farm in the provision of services, and a large part of the farms in the production of raw materials for industry (7.69%), production of animal feed (5.26%), and the use of manure, sludge, etc. (6.45%).

Public intervention (support) is required when private and market forms cannot fully govern the supply of certain agro-ecosystem services due to public nature, low appropriability, high specificity and uncertainty, etc. Participation in a public program is a form that is applied most by farms in the Fight against climate change (8.69%), Landscape and scenery protection (6.06%), and Preservation of breeds, varieties, products, etc. (4.17%).

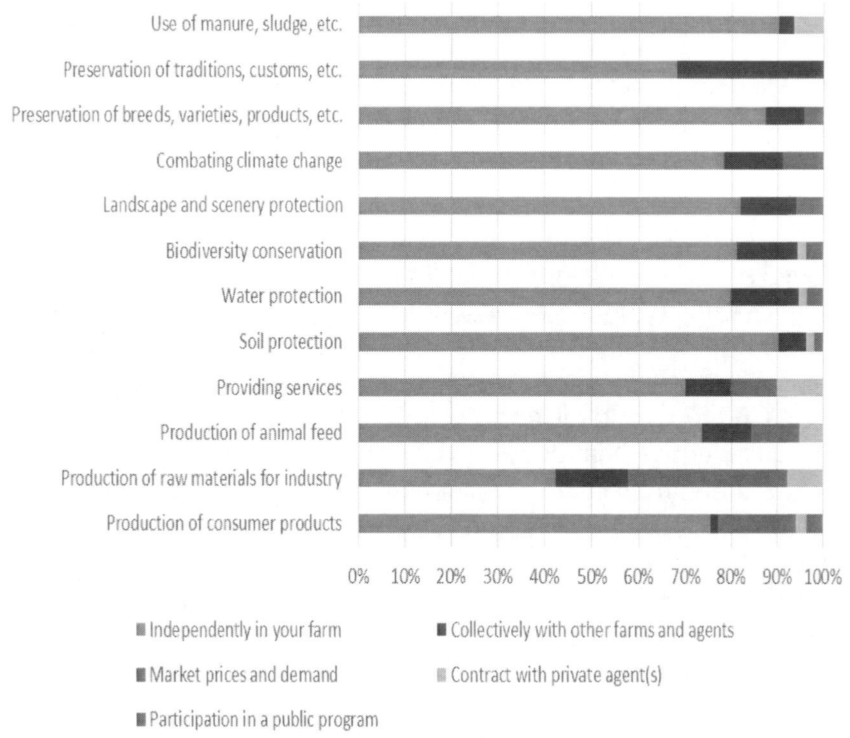

Source: Survey of agricultural producers, 2020.

Figure 10. Mechanisms used in decision-making on farm activities related to different types of agro-ecosystem services in Bulgaria.

Depending on the specificity of production (and the production agro-ecosystem), farms with different specializations use to unlike extent specific mechanisms for deciding on the activity related to agroecosystem services of different types (Figure 11). The largest share of farms specialized in Field crops (28.57%) use specific mechanisms in the production of raw materials for industry. The most widespread special mechanisms for the production of animal feed are practiced at mixed crop-livestock holdings (40.74%). Every third producer in Pigs, Poultry and Rabbits applies similar mechanisms for (standard) services provision. A significant part of the specialized in Permanent crops (43.86%) and Mix crops (36.36%) need special management mechanisms for soil Protection. In water protection, most of the holdings in Permanent crops (40.35%), Mix crop-livestock (37.04%) and Mix crops (36.36%) adapt special forms.

Farms in Permanent crops (38.60%), Mixed Livestock (38.46%), and Mixed crop-livestock (37.04%) use the most specific mechanisms for biodiversity conservation. One-third of the specialized holdings in Pigs, Poultry and Rabbits apply special forms for landscape and scenery protection. The largest part of the farms with Mix crops (27.27%) and Grazing livestock (18.18%) apply special management mechanisms in the fight against climate change. For the preservation of breeds, varieties, products, etc. and for the preservation of traditions, customs, etc. every third farm with pigs, poultry and rabbits needs such mechanisms. The majority of those specialized in Pigs, Poultry and Rabbits (66.67%) and mixed crops (63.64%) apply special mechanisms in making management decisions for the use of manure, sludge, etc.

At the same time, however, there is a significant variation in the type of specific mechanisms used to make management decisions by farms with different specializations. For example, for the Conservation of Natural Biodiversity, every third farm specializing in field crops applies Participation in a public program. When managing the supply of the same ecosystem service, two-thirds of the farms with bee colonies and one-third of those in mixed crops do it Collectively with other farms and agents. Similarly, when managing the fight against climate change, half of the Mixed Crop-Livestock holdings do so collectively with other farmers and agents, while one-fifth of the farms specializing in Permanent crops use Participation in a public program.

For some agroecosystem services with a high (capacity, location, product, etc.) specificity to a particular buyer(s) no (free)market forms (Soils protection, Waters protection, Protection of biodiversity, Preservation of landscape and scenery, Combating climate change, Preservation of breeds, varieties, products, etc.) or public forms

(Production of raw materials for industry, Production of animal feed, and Services supply), or both market and trilateral with public involvement forms (Preservation of traditions, customs, etc., and Use of manure, sludge, etc.) develop.

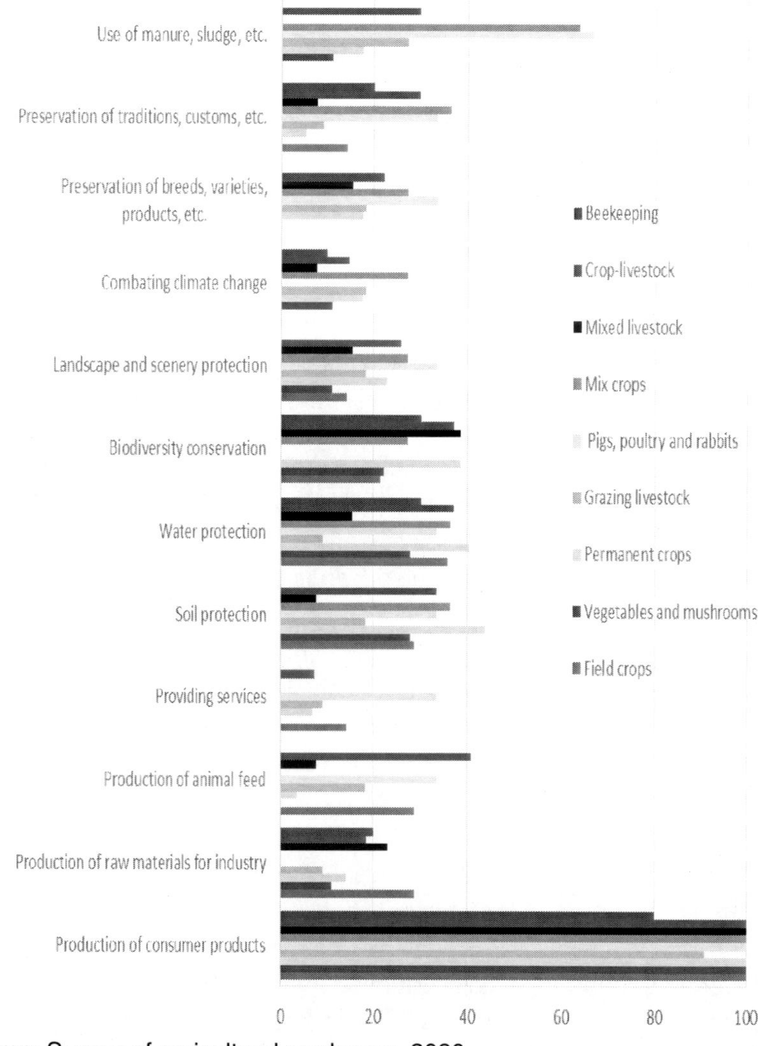

Source: Survey of agricultural producers, 2020.

Figure 11. Share of farms with different specialization, using specific mechanisms in decision-making on the activity related to agroecosystem services in Bulgaria (percentages).

For the later mostly or exclusively private (internal, contract, collective, etc.) modes are used by all types of farms to govern their activity and relations associated with ecosystem services.

Our study has found no significant differences found in specific modes of management of specific agro-ecosystem services applied by farms of different juridical types (Sole Trader, Cooperative, etc.), in different ecosystems (mountainous, plain, etc.) and regions of the country. Thus differentiation of the managerial modes mostly depends on the specificity of the agroecosystem services and the subsector of agricultural production.

4. Private, Collective and Market Modes

Most of the surveyed farms apply special private and market forms to govern the supply of agro-ecosystem services. Over 17% of all farms (17.28%) are certified for organic production, and a small part combines mixed organic and traditional production (3.09%) (Figure 12). Formal certification is associated with additional costs for farmers (conversion period, certification, current control, etc.) and consumers (premium to market price), but also brings significant benefits for both parties. Farmers have a formal guarantee for the authenticity of their products, receive a price bonus and public subsidies, develop a reputation and market position for special and high-quality products. Consumers receive a guarantee of authenticity and low-cost acquisition of products related to agri-ecosystem services. The process is controlled by an independent (third) party, which increases trust and reduces transaction costs. This three lateral market-oriented form will become even more important in the future given the growing consumer demand in the country and on international markets, and the further greening of the CAP in the next programming period and increasing incentives to expand organic production in the EU.

Most of the agricultural holdings have a built Reputation for ecologically clean products (14.81%) or with naturally ecologically clean production (19.14%).

Informal private and collective forms such as building a "good reputation" for special quality, products, origins, etc., of certain farms, ecosystems and entire regions are widespread in the country's agricultural practice. In the future, they will continue to effectively manage the relationship between producers and consumers for the supply of agri-ecosystem services. Transaction costs are low, as long-term "personal" relationships ("clientalization," high frequency) are developed for trading certain products, primarily in local and regional markets, and opportunism is punished by the cessation of trade and "bad" reputation.

Due to high costs (registrations, control, etc.) and low returns, very few farms apply other formal private or collective forms of agri-ecosystem services management. A little over 5% are members of a collective organization (5.56%), a little over 1% are With own trademark, protected origin, etc. (1.23%), less than 1% participate in a Collective Trademark, Protected Origin, etc. (0.62%) or in a Collective Initiative (0.62%).

However, given the significant transactional benefits (sales to large retail chains, exports, premiums, etc.), the number of farms investing in such special private and market forms is gradually increasing. In the process of certification are 3.01% of all farms are, with a plan for bio-certification 1.8% and With a plan for eco-brand, protected origin, etc. 1.85%.

Nearly three-quarters of the surveyed farms reported that they participate in some initiative for the protection of ecosystems and ecosystem services. The majority of farms Implement own (private) initiative in this regard (56.56%) (Figure 13). Quite a part of the holdings Implements informal Initiatives of other farms (13.11%).

Almost every tenth (9.84%) reports participating in a State initiative related to the protection of ecosystems and ecosystem services. This hybrid (public-private, trilateral) form is also usually associated with receiving certain subsidies or other support in return for certain commitments for improved environmental management. Just over 2% of farms have a contract with the state to implement such an initiative (2.46%).

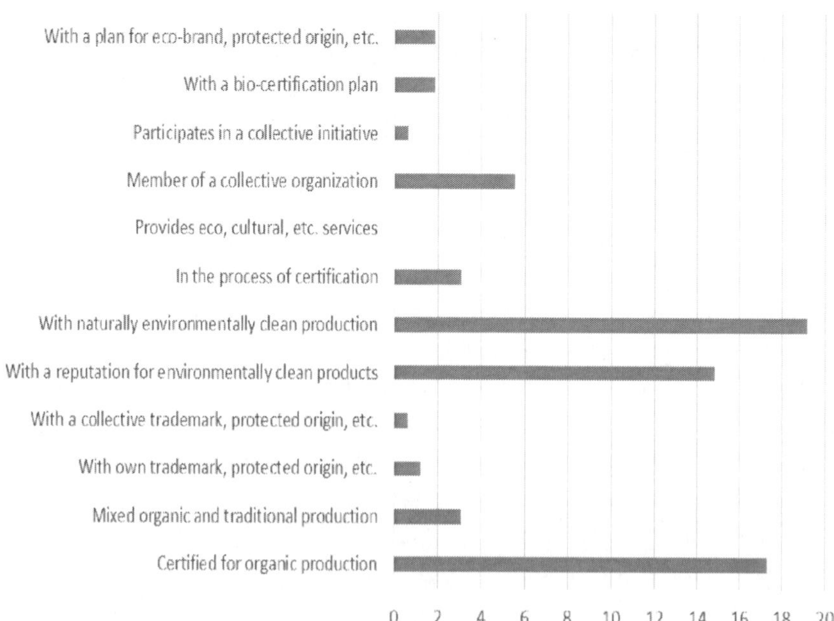

Source: Survey of agricultural producers, 2020.

Figure 12. Share of farms applying diverse private, collective, and market forms for the supply of agro-ecosystem services in Bulgaria (percentages).

A small share of farms participates in other private and collective formal environmental management initiatives - Formal initiatives of other farms (2.46%), Initiative of a professional organization (4.1%), Initiative of a non-governmental organization (3.28%), Initiative of a cooperative of which they are members (2.46%), and International initiative (0.82%).

For a small part of the farms, the initiative is of (induced by) Supplier of the farm (1.64%) or by Buyer (0.82%), and 1.64% of the farms even Have a contract with a private organization for implementation of eco-initiative.

All this shows that the effective forms that farms and other stakeholders use to govern their relationships and actions related to environmental protection and agri-ecosystem services are diversifying.

Source: Survey of agricultural producers, 2020.

Figure 13. Share of farms participating in an initiative for the protection of ecosystems and ecosystem services in Bulgaria (percentages).

5. Providing Outside Access to the Territory of the Farm

Providing external access to the territory of agricultural holdings is a basic form of supply and/or consumption of ecosystem services in agriculture.

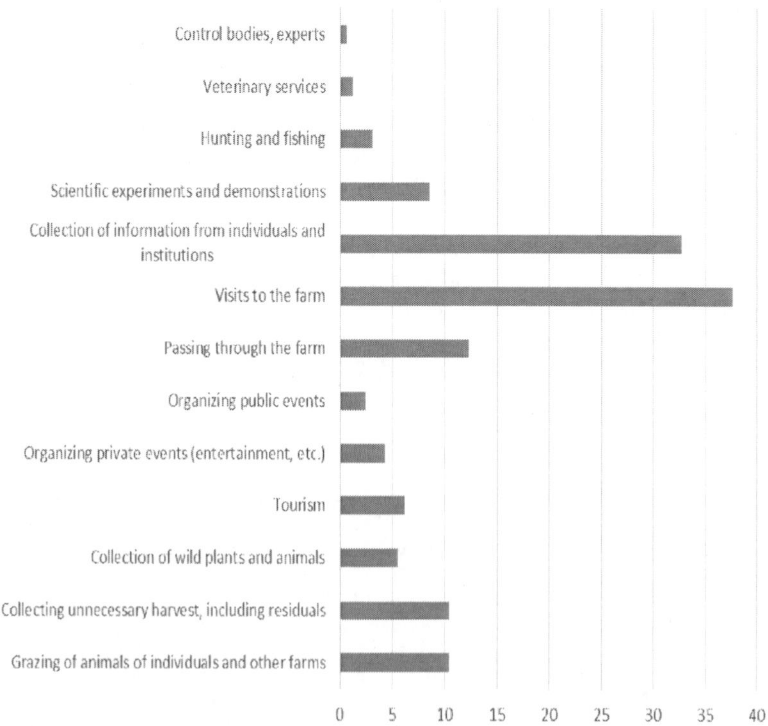

Source: Survey of agricultural producers, 2020.

Figure 14. Share of farms that provide external access to their territory for using of various ecosystem services in Bulgaria (percentages).

The share of farms that provide access to outsiders on their territory varies depending on the agroecosystem services used (Figure 14). A significant part of the farms allows External visits to the farm (37.65%) and Collection of information from individuals and institutions (32.72%). Relatively smaller is the number of farms that allow Passage through the farm (12.35%). Every tenth farm allows Grazing of animals of other individuals and farms (10.49%) and Collection of unnecessary for the farm harvest, including residues (10.49%). Quite a few of the Bulgarian farms also provide their territory for scientific experiments and demonstrations (8.64%), Tourism (6.17%) and Collection of wild plants and animals (5.55%). To the least extent, the territory of the farms is available for the Organization of private events (entertainment, etc.) (4.32%), Hunting and fishing (3.09%) and Organization of public events (2.47%). An insignificant part of the holdings also indicated other reasons such as Veterinary services (1.23%) and Control bodies and experts (0.62%).

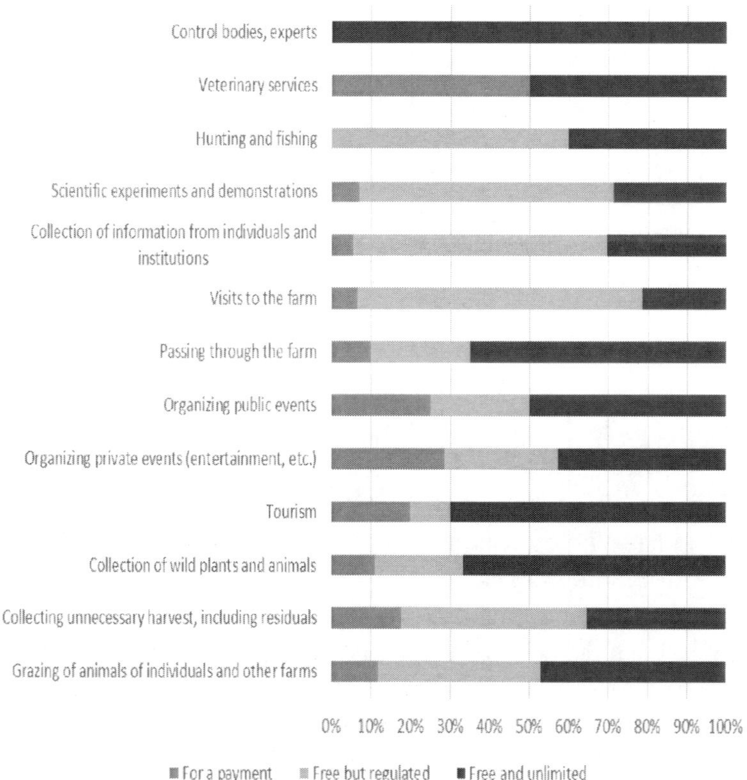

Source: Survey of agricultural producers, 2020.

Figure 15. Type of external access to farm's territory for use of different ecosystem services in Bulgaria.

For the different types of external access on the territory of the farms, specific forms for governing the relationship of agents are practiced (Figure 15). Free and unrestricted access is the dominant form of providing access to the territory of the farm for grazing animals of individuals and other farms (47.06%), Collection of wild plants and animals (66.67%), Tourism (70%), Organizing private events (42.86%), Organization of public events (50%), Passage through the farm (65%), Veterinary services (50%) and Control bodies and experts (100%). This form is also practiced by a large number of farms for the Collection of unnecessary harvest, residues (35.29%), Collection of information from individuals and institutions (30.19%), Scientific experiments and demonstrations (28.57%), Visits to the farm (21.31%), and Hunting and fishing (40%). All these agro-ecosystem services are treated as public goods and their use and consumption are "managed" by providing free and unrestricted access by farm owners. Most of these services are difficult to regulate or exchange as private goods due to high uncertainty and enforcement costs.

In many cases the main form for providing access to the territory for the farm is Free but regulated - for Collection of unnecessary crops, residues (47.06%), Visits to the farm (72.13%), Collection of information from individuals and institutions (64.15%), Scientific experiments and demonstrations (64.28%) and Hunting and fishing (60%). This form is widely used by a large number of farms in allowing access to the territory for Grazing animals of individuals and other farms (41.18%), Collection of wild plants and animals (22.22%), Organization of private events (28.57%), organizing public events (25%) and Passing through the farm (25%). The use and consumption of this type of agro-ecosystem services are managed through a private form - regulation, and they are provided free of charge by farm owners. The form of free provision is determined either by the additional benefits received for the farmers (in case of grazing animals of individuals and other farms, collection of unnecessary crops, residues, collection of wild plants and animals, organization of private and public events, etc.), or from the high costs of enforcement - constant control, penalties, disputing through a third party, etc. (in Passing through the territory of the farm, Hunting and fishing, etc.). Here, regulation is needed to plan and coordinate external access and/or limit consumption to maintain a sustainable supply of agro-ecosystem services.

A portion of farms uses a market form of exchange against payment of a price to provide external access to the territory of the farms. This form of sale of services is practiced in grazing animals on individuals and other farms (11.76%), collection of unnecessary crops, residues (17.65%), collection of wild plants and animals (11.11%), tourism (20%), organizing private events (28.57%), organizing public events (25%), passing through the farm (10%), visits to the farm (6.56%), gathering information from individuals and institutions (5.66%), scientific experiments and demonstrations (7.14%) and veterinary Services (50%). The market form is preferred because it governs well the supply of "limited" agro-ecosystem services and relationships of counterparts. Market trading is beneficial for both parties, who mutually profit from the transaction, as the terms of exchange are easy for no or low-cost negotiation, control and sanctioning. Here, the classic contract of "spot like" exchange under standard conditions applies, and payment is made on the spot or in advance to avoid any possible opportunism.

Agricultural holdings with different specializations provide unequal external access on the territory to farms for using different agro-ecosystem services (Figure 16). To the greatest extent outside access to the territory of the farm for grazing animals of individuals and other farms is provided by holdings specialized in Grazing livestock (36.36%) and mixed crop-livestock operations (22.22%).

For Harvesting of unnecessary output, incl. Residues, most farms providing external access to their territory are among specialized in field crops (21.43%) and crop-livestock (14.81%). The largest share of mix crop-livestock farms (11.11%) also allows the collection of wild plants and animals and tourism in their territory.

Specialized in grazing livestock to the greatest extent provide external access on the territory of their farms for Organizing private events (entertainment, etc.) (18.18%) and Organizing public events (9.09%).

Most farms that allow passage through the farm territory are among those specialized in permanent crops (19.30%) and grazing animals (18.18%). Most visits to the farm are allowed by farms specializing in grazing animals (63.64%) and field crops (50%).

The largest share of farms that allow the collection of information from individuals and institutions are among those specializing in permanent crops (43.86%) and grazing animals (36.36%), and for scientific experiments and demonstrations among those specializing in grazing animals (27.27%) and Bee families (20%). Every tenth farm with bee families also allows the use of its territory for hunting and fishing.

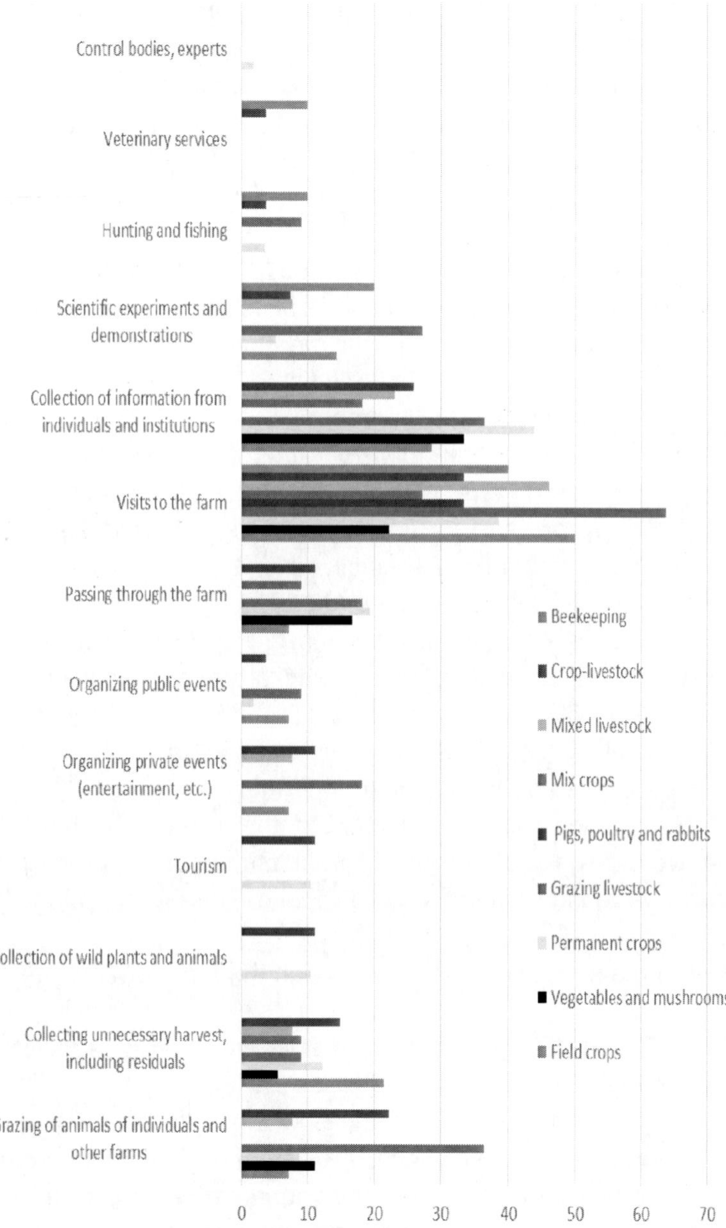

Source: Survey of agricultural producers, 2020.

Figure 16. Share of farms with a different specialization that provides external access to their territory for use of agro-ecosystem services in Bulgaria (percentages).

Therefore, in addition to the product specialization, there is a certain specialization in the provision of agro-ecosystem services related to external access on the territory of the farms.

Farms with different specializations use unequally different forms for ensuring open access to the territory of farms for the use of agro-ecosystem services. The preferred most efficient mode is (pre)determined by the specifics of the production and the use of territory and/or the preferences of the owners/managers of the individual farms and the external users of the related agro-ecosystem services. For example, for farms specialized in field crops, vegetables and mushrooms, and mixed livestock, Free but regulated access is the only form used for providing external access to the territory for grazing animals to individuals and other farms. At the same time, most of the farms specializing in permanent crops practice Free and Unrestricted Access, while the remaining one-fifth apply for Paid access.

Similarly, relations with clients associated with harvesting unnecessary output, incl. residues on the territory of farms specialized in Vegetables and Mushrooms, Grazing livestock and Mixed crops are managed entirely on a contractual basis for payment. At the same time, for all other groups of farms, the used form is either Free but regulated or Free and unrestricted access.

6. Efficiency and Importance of Farms' Ecosystem Services Provision

According to the majority of managers of the surveyed farms, their activity for the protection of ecosystems and their services is associated with an Increase in the total production costs of the farm, Increase of the specialized costs for nature protection, Increase of long-term investments, Increase of management costs and efforts, Growth of the costs of participation in state aid programs, Increase in the costs of studying the regulations and standards, and Increase in the costs of registrations, tests, certification, etc. (Figure 17). Moreover, for the majority of farms this activity leads to a high increase in the total production costs of the farm (50%), the specialized costs for nature protection (40.58%), long-term investments (50.7%), the costs for participation in state aid programs (40%), and the costs of registrations, tests, certification, etc. (50.75%). At the same time, for only a small part of all farms, environmentally-friendly activity is associated with a reduction in the various types of production and transaction costs.

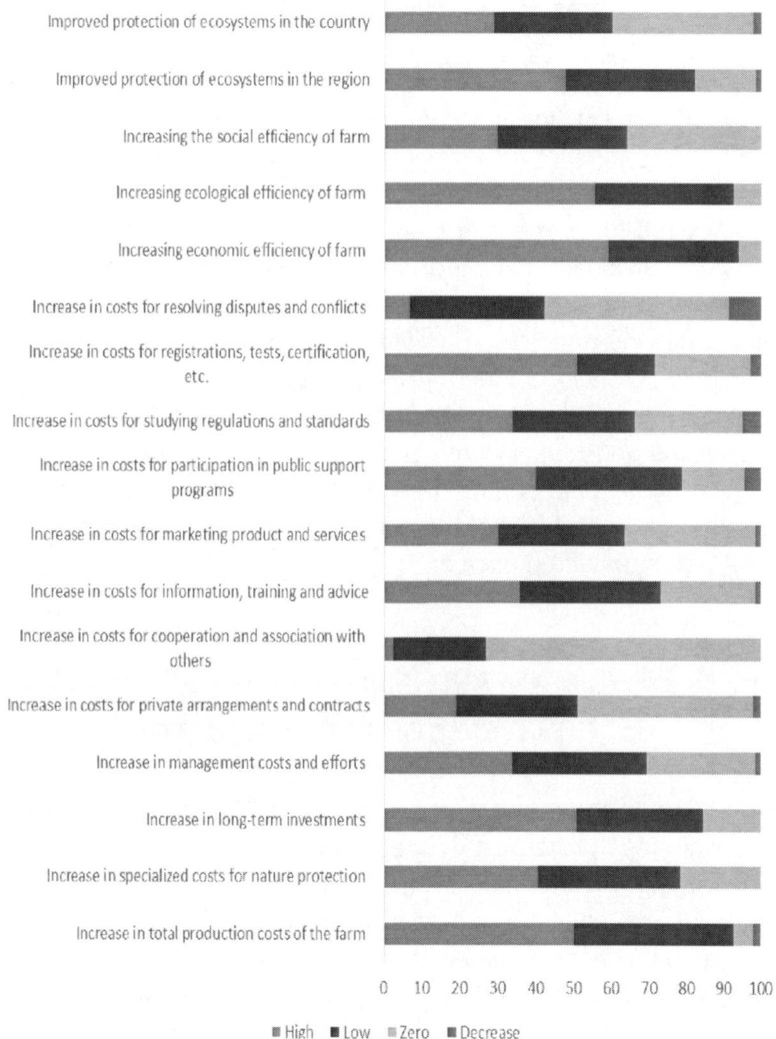

Source: Survey of agricultural producers, 2020.

Figure 17. Costs and efficiency of the activity of farms for protection of ecosystems and their services in Bulgaria (percentages).

At the same time, however, the vast majority of farms report that their activities for the protection of ecosystems and their services are also associated with an Increasing the economic efficiency of the farm, Increasing the ecological efficiency of the farm, Increasing the social efficiency of the farm, Improved protection of ecosystems in the region, and Improved protection of ecosystems in the country. At the same time, the majority of farms estimate that their environmentally friendly activity leads to a high increase in the economic efficiency of the farm (59.09%), the ecological efficiency of the farm (55.22%) and the protection of ecosystems in the region (47.54%).

None or very few of the surveyed farms indicate that their activities for the protection of ecosystems and their services are related to reducing the economic efficiency, environmental and social efficiency of the farm, and the protection of ecosystems in the region and the country. However, a significant share of farm managers believes that their efforts and costs to protect ecosystems and ecosystem services do not lead to changes in the social efficiency of the farm (36.17%) and improved protection of ecosystems in the country (37.78%).

There is significant differentiation in the level of costs and efficiency of farm activities related to the protection of ecosystems and ecosystem services (Figure 18). A high increase in the total production costs of the farm was reported by half of the farms specializing in field crops and mixed crop production, three-quarters of those in grazing animals, and all of those in bee colonies. The share of farms with a high increase in these costs is the smallest among holdings specialized in vegetables and mushrooms (every third) and none in pigs, poultry and rabbits.

The largest share of farms with a high increase in specialized costs for nature protection are among those specialized in field crops, mixed crop production and crop and mix crop-livestock production (50% each) and bee families (100%). At the same time, relatively few mixed livestock farms (20%) reported a high increase in this type of cost, and none among those specializing in grazing animals and pigs, poultry and rabbits.

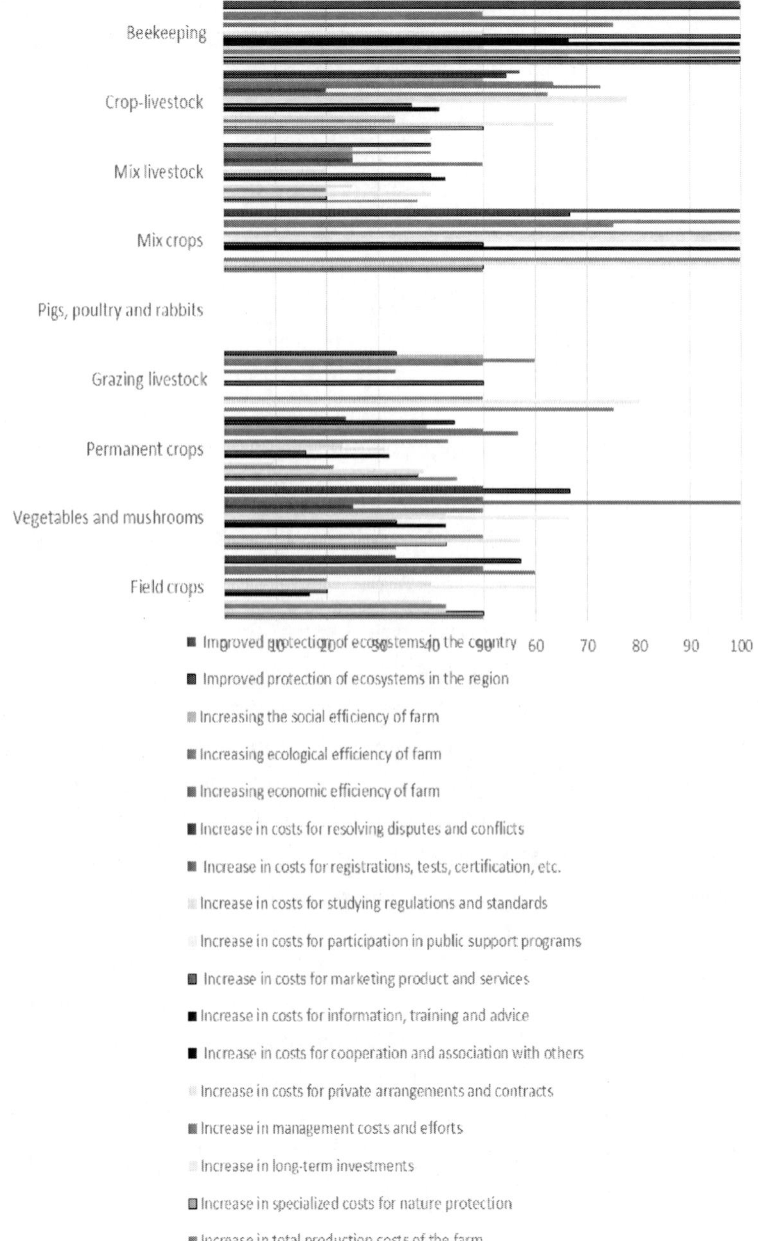

Source: Survey of agricultural producers, 2020.

Figure 18. Share of farms with a *high* increase in costs and efficiency of activity for the protection of ecosystems and their services in Bulgaria (percentages).

A high Increase in long-term investments for the protection of ecosystems and ecosystem services is most typical for farms specializing in Vegetables and mushrooms (57.14%), Herbivores (80%), Mixed crop production (100%), Crop and livestock production (63.64%) and Bee families (66.67%). The lowest share of farms with high costs of this type is in Permanent crops (38.71%), and in none of the surveyed farms in Pigs, poultry and rabbits.

High increases in management costs and efforts to protect ecosystems and ecosystem services are recorded in most of the farms specializing in Vegetables and Mushrooms and Herbivores (every second of them) and Mixed crop production and Bee Families (all). At the same time, relatively few of the farms in Perennials (21.4%) and Mixed Livestock (20%), and none of those in Pigs, Poultry and Rabbits reported a high increase in these costs.

For a high increase in the costs of private arrangements and contracts related to the protection of ecosystems and ecosystem services, most farms report in Field Crops (40%) and Bee Families (50%), while in other groups a small number or none of the holdings have growth in these costs.

A high increase in the costs of cooperation and association with others related to the protection of ecosystems and ecosystem services is observed in all farms specializing in beekeeping, while in other categories of farms this type of cost is not typical.

The most numerous are the farms with high Increase in costs for information, training and advice on ecosystem protection and ecosystem services in those specialized in Mixed Crop Production (100%) and Bee Families (66.67%), and relatively few in Field Crops (16.67%) and none for Grazing animals, and Pigs, poultry and rabbits.

The largest share of farms with a high increase in the cost of marketing the product and services related to the protection of ecosystems and ecosystem services is in those specializing in grazing animals and mixed crop production (every second of them), bee families (all), relatively few in field crops (20%) and perennials (16%) and none among those in pigs, poultry and rabbits.

Most of the farms report high growth in the costs of participation in state aid programs related to the protection of ecosystems and ecosystem services, among those specialized in field crops (60%), vegetables and mushrooms (66.67%), mixed crop production (100%), and mix crop-livestock (77.78%). On the other hand, relatively fewer farms reported similar growth among specialized in perennials (31.03%) and mixed livestock (20%) and none of those with grazing animals and pigs, poultry and rabbits.

The high growth of expenditures for studying regulations and standards related to the protection of ecosystems and ecosystem services was noted by the largest number of farms with Mixed crop produces (100%) and Crop-livestock specialization (77.78%). At the same time, a relatively small proportion of farms specializing in perennials (23.08%) and none of those in grazing animals, pigs, poultry and rabbits, mixed livestock and bee colonies reported a similar increase in this type of expenditure.

The high growth of expenditures for registrations, tests, certification, etc. related to the protection of ecosystems and ecosystem services is observed in most farms with Mixed Crop Production (100%), Crop-Livestock production (62.5%) and Bee Families (75%). This share is lowest on farms in field crops (20%) and on none of those in pigs, poultry and rabbits.

High growth in the costs of resolving disputes and conflicts related to the protection of ecosystems and ecosystem services is reported by every fourth farm specializing in Vegetables and Mushrooms and Mixed Livestock and every fifth of those in Bee colonies. However, none of the other holdings reported a similar increase in this type of expenditure.

High increase of the economic efficiency of the farm-related to the protection of ecosystems and ecosystem services is most noted in the farms specialized in Field crops (60%), Vegetables and mushrooms (100%), Mixed crop production (75%), Mix crop-livestock production (72.73%) and Bee families (100%), and the least in those in Mixed livestock (25%) and Pigs, poultry and rabbits (0).

A high increase of the ecological efficiency of the holdings' activity for the protection of ecosystems and ecosystem services is reported by all from Mixed crops farms, and the majority of those with Grazing animals (60%) and Crop and animal husbandry (63.64%). The lowest share of farms with similar growth is in those specialized in Mixed Livestock (40%) and Pigs, poultry and rabbits (0).

High Increasing the social efficiency of the holdings' activity for the protection of ecosystems and ecosystem services is registered by every second farm specializing in Herbivores and Corp-livestock, a smaller part of those in Perennial crops (39.13%) and Mixed livestock (25%), and from none of the other categories of holdings.

High improved protection of ecosystems in the region, related to the activity of farms for protection of ecosystems and ecosystem services is achieved mostly by the farms in Field crops (57.14%), Vegetables and mushrooms (66.67%), Mixed crop growing (66.67%), and Bee families (100%), and relatively the least of those with Grazing animals (33.33%) and Pigs, poultry and rabbits (0).

High improved protection of ecosystems in the country related to the activities of farms for protection of ecosystems and ecosystem services is reported by all those specializing in Mixed crops and Bee families, and most of those in Mix crop-animal husbandry (57.14%). The share of farms with a similar effect is the lowest in those specialized in field crops (33.33%) and perennials (23.81%), and in none of them in grazing animals, pigs, poultry and rabbits, and mixed animal husbandry.

The vast majority of farm managers estimate that the effect of the overall activity of the farm is positive in terms of soils (73.95%), biodiversity (62.3%), landscape (51.11%) and economic development of the region (60.82%) (Figure 19). Also, the majority of managers believe that the effect is positive in terms of Air (48.54%), Surfacewaters (36.2%), Groundwaters (47.47%), Climate (38.37%), Traditional breeds, varieties, products, technologies. (44.68%), and Social development of the region (48.89%), as a relatively smaller part consider a positive effect in terms of Local culture, traditions, customs, education (28.39%).

However, the share of managers who believe that the whole activity of their farm is not associated with any effect on the individual elements of the ecosystem - Soils (14.29%), Air (29.13%), Surfacewaters (34%), Groundwaters (26.26%), Biodiversity (16%), Landscape (17.78%), Climate (23.26%), Traditional breeds, varieties, products, technologies (20.21%), Local culture, traditions, customs, education (32.1%), Economic development of the region (16.49%) and Social development of the region (18.89%).

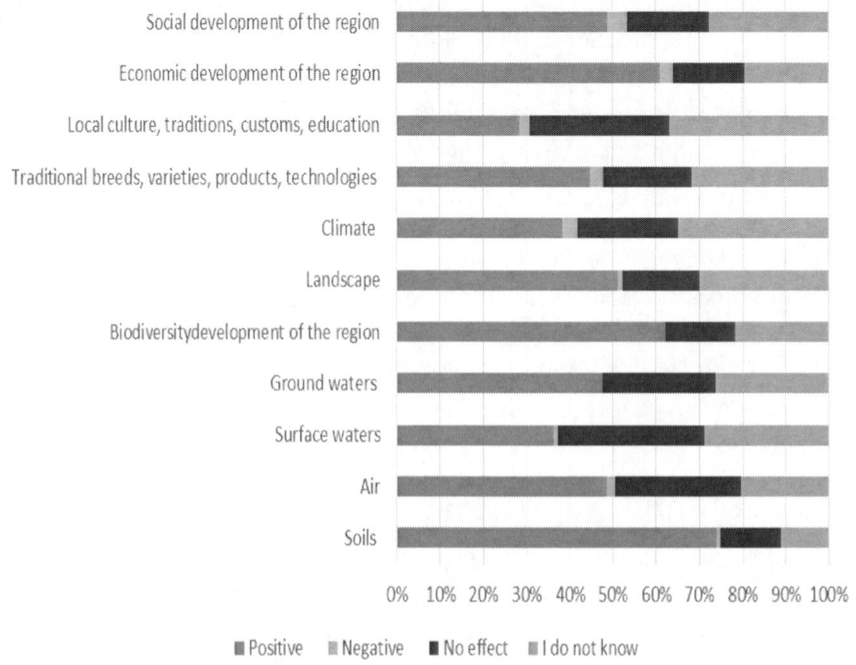

Source: Survey of agricultural producers, 2020.

Figure 19. Effect of farms' overall activity on different elements of ecosystems and their services in Bulgaria.

In addition, a significant part of managers do not know the effect of the overall activity of agriculture on various elements of the ecosystem - Soils (10.92%), Air (20.39%), Surfacewaters (28.7%), Groundwaters (26.26%), Biodiversity (21.7%), Landscape (30%), Climate (34.88%), Traditional breeds, varieties, products, technologies (31.91%), Local culture, traditions, customs, educated (37.04%), Economic development of the region (19.59%), and Social development of the region (27.78%). The later requires both deepening and expanding independent assessments of the effects of farming on the individual components of ecosystems, and better informing farmers about their negative and/or positive contribution to environmental protection and ecosystem services.

Just over half of the surveyed managers assess the importance of their activities for the protection of agro-ecosystems and agro-ecosystem services as High for their farm (50.62%) and 46.91% High for themselves (Figure 20). A significant share of managers also believes that their activities for the protection of agro-ecosystems and agro-ecosystem services are of high importance for the region of their farm (27.16%). There is also a significant number of managers who believe that this activity has a high environmental value (14.81%) and value for future generations (13.58%). A relatively smaller part of the managers believes that such activity is of High importance for the community in the region (7.41%), High market value (5.56%) and High economic value (6.17%).

At the same time, an insignificant share of managers is convinced that their activity for the protection of agro-ecosystems and agro-ecosystem services has a High contract value (1.23%), and a High social value (2.47%) or is Without any value (1.23%), as none of the respondents believes that this activity has a High cultural value.

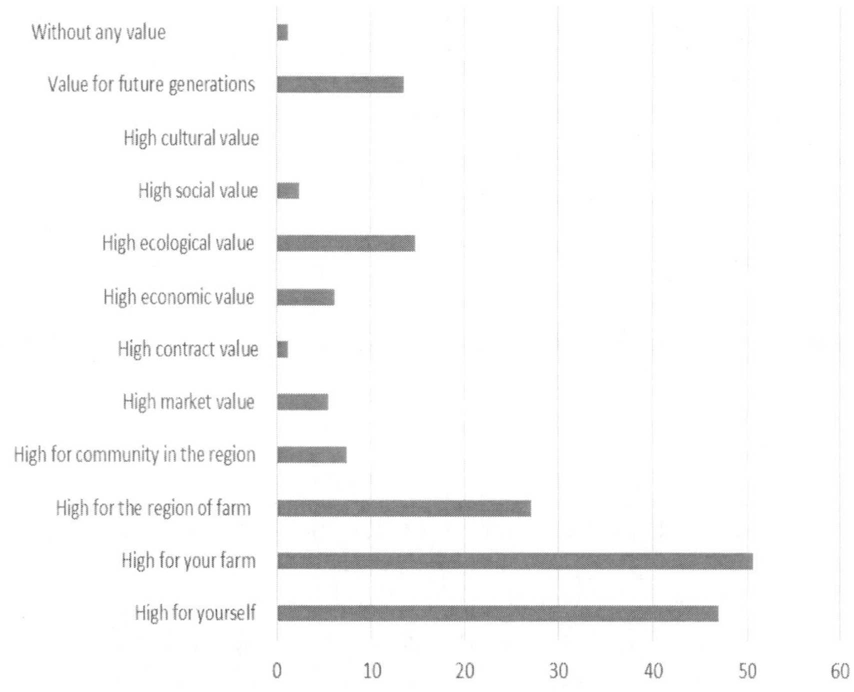

Source: Survey of agricultural producers, 2020.

Figure 20. Assessment of farm managers of the importance of their activity for the protection of agro-ecosystems and agro-ecosystem services in Bulgaria (percentages).

7. Factors in the Governance of Agro-Ecosystem Services

The survey allows us to identify personal, organizational, market, institutional and other factors that have the greatest impact on (and predetermine) the activity of agricultural holdings for the conservation of agro-ecosystems and agro-ecosystem services.

According to the majority of surveyed managers, the factors that strongly stimulate or limit the activity of farms related to the preservation of agro-ecosystems are Market demand and prices (51.23%), Market competition (37.65%), Opportunities to increase profits (37.65%), Participation in state and European support programs (37.04%), Financial capabilities (35.8%), Direct state and European subsidies received (34.57%), Personal conviction and satisfaction with this activity (30.86%), Amount of direct costs for this activity (29.63%), Access to farmers' advice (24.69%), Regulatory documents, standards, norms, etc. (24.69%) and State Policy (23.46%) (Figure 21).

The extent to which the activity for the protection of the agroecosystems of the affected farms is stimulated or limited by different factors is not the same. Factors that *strongly stimulate* the activity of the majority of agricultural producers for protection of agro-ecosystems and their services are: Market demand and prices (69.88%), Market competition (57.38%), Opportunities to increase profits (78.69%), Initiatives and pressure of the public and interest groups (61.11%), The presence of cooperation partners in this activity (55%), Private contracts for the sale of related products and services (65%), Initiatives of other farms (68.18%), Immediate benefits for the farm in present and near future (82.76%), Long-term benefits for the farm (86.21%), Benefits for others (75%), Integration with the supplier of the farm (81.25%), Integration with the buyer of the production (80.95%), Integration with processor (80%), Available information and innovation (91.3%), Professional training of managers and employees (91.67%), Access to farmers' advices (92.5%), Received direct state and European subsidies (91.07%), Participation in state and European support programs (95%), Tax preferences (67.86%), Existence of a long-term contract with the state (68.42%), Positive experience of other farms and organizations (87.5%), Policies of the European Union

(68.96%), Public recognition of contribution (60.87%), and Personal conviction and satisfaction with this activity (88%) (Figure 22).

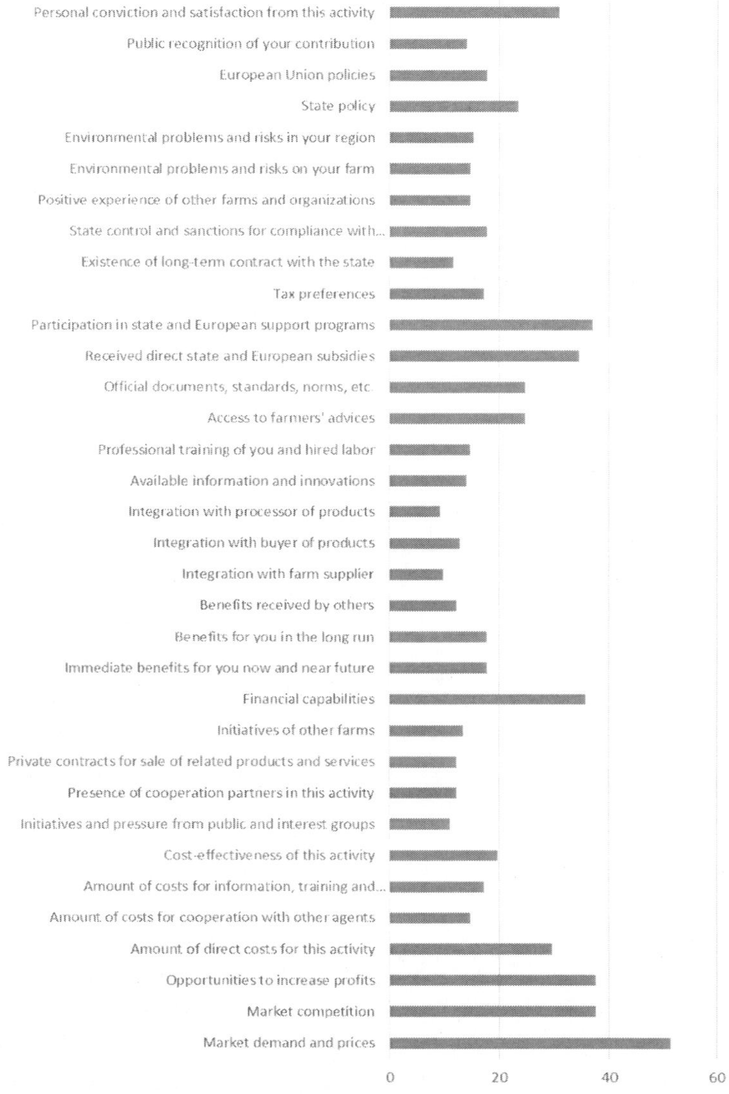

Source: Survey of agricultural producers, 2020.

Figure 21. Factors that strongly stimulate or restrict the activity of farms related to conservation of agroecosystems in Bulgaria (percentages).

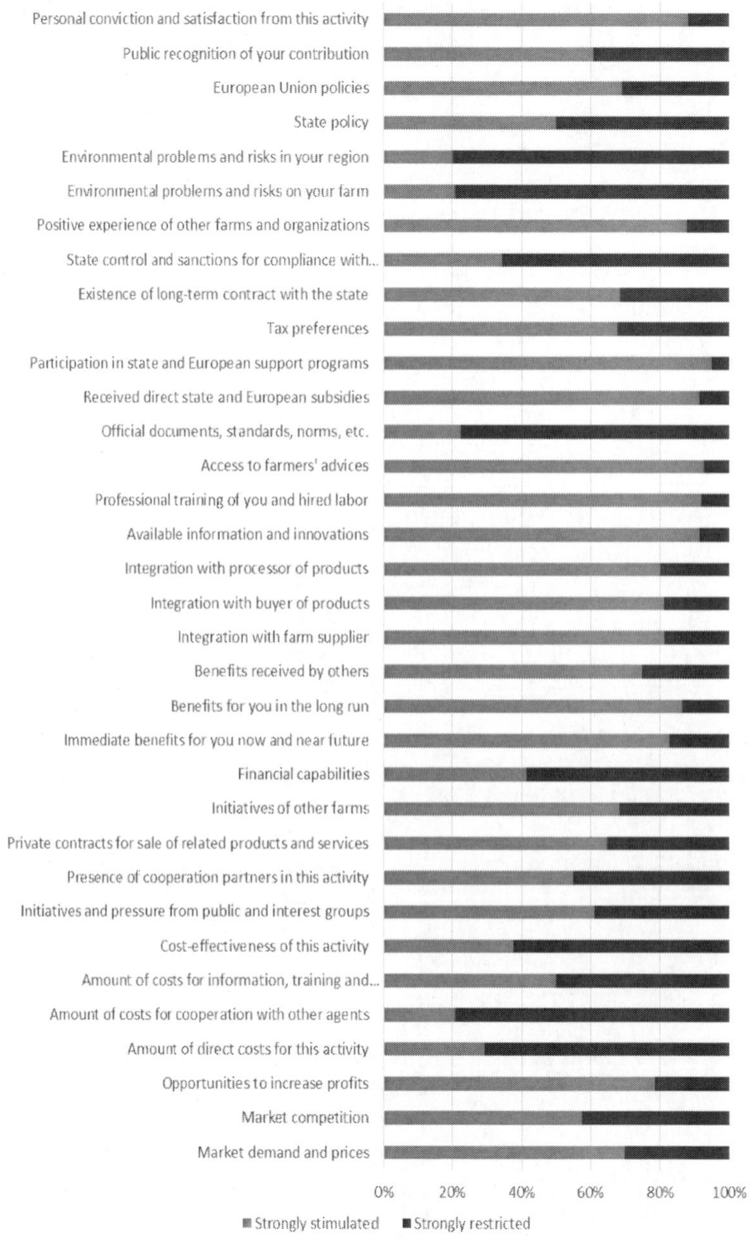

Source: Survey of agricultural producers, 2020.

Figure 22. The extent to which farming activities related to the conservation of agroecosystems are stimulated or limited by various factors in Bulgaria (percentages).

Factors that severely limit the activity of the majority of farms for the protection of agro-ecosystems and their services are the Amount of direct costs for this activity (70.83%), the Amount of costs for cooperation with other agents (79.17%), Economic efficiency of costs for this activity (62.5%), Financial capabilities (58.62%), Regulatory documents, standards, norms, etc. (77.5%), State control and sanctions for compliance with standards, norms, etc. (65.52%), Environmental problems and risks in the farm (79.17%) and Environmental problems and risks in the region (80%).

At the same time, the Amount of information, training and consultation costs, and the State Policy are factors that strongly stimulate the environmentally friendly activity of half of the surveyed farms, and severely limit it for the other half.

All these factors are to be taken into account when improving public policies and forms of intervention related to the governance of agro-ecosystems and their services.

CONCLUSION

It is well known that agricultural production makes a significant contribution to the conservation, restoration and enhancement of ecosystems and their services, but also is associated with negative effect and their degradation and demolition („agricultural disservices“). Therefore, services related to agricultural production and agro-ecosystems are among the most intensively studied, mapped, evaluated, regulated and stimulated.

Our study has tried to fill the gap and give initial insights on great variety of agricultural services and their importance for the farm, region, other ecosystems and agents in Bulgaria. At the current stage of development country's farms maintain or provide a great number of essential ecosystem services among which provisioning food and feed, and conservation of elements of the natural environment prevailing. Besides, there are significant differences in the participation and contribution of agricultural holdings in the protection and provision of agro-ecosystem services in the various specific and principled ecosystems of the country, and major subsectors of agricultural production. The later requires special measures to improve, diversify and intensify this activity of farmers through training, information, exchange of experience, public incentives and support, etc.

The study has also found out that there is significant differentiation of employed managerial forms depending on the type of ecosystem services and specialization of agricultural holdings. Management of agroecosystem services is associated with a considerable increase in the production and transaction costs of participating farms as well as big socio-economic and environmental effects for holdings and other parties. Factors that mostly stimulate the activity of Bulgarian producers for protection of agro-ecosystems and their services are participation in public support programs, access to farmers' advice, professional training, available information and innovation, received direct subsidies, personal conviction and satisfaction, positive experience of others, long-term and immediate benefits for the farm, and integration with suppliers, buyers and processors.

Suggested holistic and interdisciplinary framework for analyzing the structure and management of agro-ecosystem services is to be extended and improved, and widely and periodically applied in the future. The latter requires systematic in-depth multidisciplinary research in this new area, as well as collection of original micro and macro-information on structure of agro-ecosystems services, and forms, efficiency and factors of agroecosystem services management by agents involved in (joint) production and management of agro-ecosystem services of a different type. The accuracy of analyzes is to be also improved by increasing representativeness through enlarging the number of surveyed farms and related agents, applying statistical methods, special "training" of implementers and participants, etc. as well as improving the official system for collecting agricultural, agro-economic and agri-environmental information in the country.

References

Adhikari B. and G. Boag (2013): Designing payments for ecosystem services schemes: some considerations, *Current Opinion in Environmental Sustainability 2*013, 5:72–77.

Allen J., J.y DuVander, I. Kubiszewski, E. Ostrom (2011): *Institutions for Managing Ecosystem Services Solutions*, Vol. 2, 6, 44-49.

Bachev H. (2009): Management of agro-ecosystem services, *Economics and management of agriculture* No 6, 3-20.

Bachev H. (2009): Governing of Agro-ecosystem Services. *Modes, Efficiency, Perspectives,* VDM Verlag.

Bachev H. (2010): *Governance of Agrarian Sustainability*, New York: Nova Science Publishers.

Bachev H (2011): Management of Agro-Ecosystem Services: Framework of Analysis, Case of Bulgaria, in J. Daniels (editor), *Advances in Environmental Research.* Vol. 17, New York: Nova Science, 119-164.

Bachev H. (2012): Governing of Agro-Ecosystem Services in Bulgaria, in A. Rezitis (editor), *Research Topics in Agricultural and Applied Economics,* Vol. 3, Bentham Science Publisher, 94-129.

Bachev H. (2012): Efficiency of farms and agricultural organizations, *Economic Thought*, Vol. 4, 46-77.

Bachev H. (2014): Eco-management in agriculture, *Economic Thought*, issue 1, 29-55.

Bashev H. (2016): Defining and assessment of sustainability of farms, *Economic Studies Journal, 1*58-188.

Bachev H. (2018): *The Sustainability of Farming Enterprises in Bulgaria*, Cambridge Scholars Publishing.

Bachev H., B. Ivanov, D. Toteva (2019): Assessment of the socio-economic and ecological sustainability of agricultural ecosystems in Bulgaria, *Economic Thought*, issue 2, 33-56.

Bachev H., B. Ivanov, D. Mitova, P. Marinov, K. Todorova, A. Mitov (2020): *Approach for Assessment of the Management of Agroecosystem Services in Bulgaria*, IAI - Sofia.

Bachev H. (2020): Approach to Analysis and Improvement of the Management of the Services of the Agro-Ecosystems, *Economics and Management of Agriculture*, issue 3, 27-48.

Bashev H. (2020): Defining, Analyzing and Improving the Management of the Services of the Agro-Ecosystems, *Economic Thought*, Vol. 4, 3-30.

Bachev H. (2020): Defining, analyzing and improving the governance of agroecosystem services, *Economic Thought,* 4, 31-55.

Bachev H. (2020): Defining, analyzing and improving the governance of agroecosystem services, *Economic Thought*, 4, 31-55.

Bachev H. (2020): Understanding and improving the governance of ecosystem services: The case of agriculture, *Journal of Economics Bibliography,* Volume 7, Issue 3, 170-195.

Bachev H. (2020): Understanding, evaluating and improving the system of governance of agro-ecosystem service, *Exploratory Environmental Science Research*, Vol. 1, Issue 1, 96-114.

Bachev H. (2020): About the Governance of Agro-ecosystem Services, *Open Journal of Economics and Commerce,* Volume 3, Issue 1, 24-36.

Bachev H. (2021): A Study on Amount and Importance of Ecosystem Servives from Bulgarian Agriculture, *Journal of Business Analytics and Data Visualization,* Volume-2, Issue-1, 7-27.

Bachev H. and D. Terziev (2019): Sustainability of Agricultural Industries in Bulgaria, *Journal of Applied Economic Sciences,* Volume 14, Issue 1.

Bachev H., B.Ivanov and A.Sarov (2020): Unpacking Governance Sustainability of Bulgarian Agriculture, *Икономически изследвания,* 6, 106-137. [*Economic research*]

Bachev H., B.Ivanov, A. Sarov (2020): *Why and How to Assess the "Governance" Aspect of Agrarian Sustainability - The Case of Bulgaria, Agricultural Research Updates.* Volume 30, Editors P. Gorawala and S. Mandhatri, Nova Science Publisher, 49-104.

Bachev H. (2021): Assessing and Improving the Governance of Agroecosystem Services, P. Gorawala, S. Mandhatri (Editors) *Agricultural Research Updates.* Volume 33, 53-92.

Bachev H., B.Ivanov and A.Sarov (2021): Assessing Governance Aspect of Agrarian Sustainability in Bulgaria, *Bulgarian Journal of Agricultural Sciences*, 2.

Boelee, E. (Editor) (2013): *Managing water and agroecosystems for food security*, CABI.

De Groot R., Wilson M, Boumans R. (2002): A typology for the description, classification and valuation of ecosystem functions goods services. *Ecol Econ* 41:393–408.

EEA (2015): *Ecosystem services in the EU, European Environment Agency.*

EEA (2020): *Ecosystems and Ecosystem Services, Executive Environment Agency (EEA).*

FAO (2016): *Mainstreaming ecosystem services and biodiversity into agricultural production and management in East Africa, Technical guidance document, FAO.*

Fremier A., F. DeClerck, N.Bosque-Pérez, N. Carmona, R, Hill, T. Joyal, L. Keesecker, P. Klos, A. Martínez-Salinas, R. Niemeyer, A. Sanfiorenzo, K. Welsh, J. Wulfhorst (2013): Understanding Spatiotemporal Lags in Ecosystem Services to Improve Incentives, *BioScience* Vol. 63 No. 6.

Gao H., T. Fu, J. Liu, H. Liang and L. Han (2018): Ecosystem Services Management Based on Differentiation and Regionalization along Vertical Gradient, China, *Sustainability,* 10, 986

Garbach K., J. Milder, M Montenegroand, F. DeClerck (2014): *Biodiversity and Ecosystem Services in Agroecosystems*, Elsevier.

Gemmill-Herren B. (2018): Pollination Services to Agriculture Sustaining and enhancing a key ecosystem service, Routledge.

Grigorova Y. & Kazakova Y. (2008): *High Nature Value farmlands: Recognizing the importance of South East European landscapes, Case study report*, Western Stara Planina, WWF (EFNCP).

Habib T., S. Heckbert, J. Wilson, A.Vandenbroeck, J. and D. Farr (2016): Impacts of land-use management on ecosystem services and biodiversity: an agent-based modelling approach. *Peer J* 4:e2814.

INRA (2017): *A framework for assessing ecosystem services from human-impacted ecosystems. EFESE.*

Kanianska R. (2019): *Agriculture and Its Impact on Land-Use, Environment, and Ecosystem Services*, INTECH.

Kazakova J. (2016): *Agriculture with high natural value (training, innovation, knowledge)*, UNWE.

Laurans Y. and L.Mermet (2014): Ecosystem services economic valuation, decision-support system or advocacy? *Ecosystem Services,* Vol. 7, 98-105.

Lescourret F., D. Magda, G. Richard, A. Adam-Blondon, M. Bardy, J. Baudry, I. Doussan, B. Dumont, F. Lefèvre, I. Litrico, R. Martin-Clouaire, B. Montuelle, S. Pellerin, M. Plantegenest, E. Tancoigne, A.Thomas, H. Guyomard, J. Soussana (2015): A social–ecological approach to managing multiple agro-ecosystem services, *Current Opinion in Environmental Sustainability,* Vol. 14, 68-75.

Marta-Pedroso C., L. Laporta, I. Gama, T. Domingos (2018): Economic valuation and mapping of Ecosystem Services in the context of protected area management, *One Ecosystem* 3: e26722,

Munang R., I. Thiaw, K. Alverson, J. Liu and Z. Han (2013): The role of ecosystem services in climate change adaptation and disaster risk reduction, *Current Opinion in Environmental Sustainability,* 5:47–52.

MEA (2005): *Millennium Ecosystem Assessment, Ecosystems and Human Well-being*, Island Press, Washington, DC.

Nedkov S. (2016): *Concept for Ecosystem Services, Presentation, workshop* May 31, 2016.

Nikolov S. (2018): Ecosystem services and their assessment - a brief overview, *Journal of the Bulgarian Geographical Society*, Volume 39, 51–54.

Novikova A., L. Rocchi, V. Vitunskienė (2017): Assessing the benefit of the agroecosystem services: Lithuanian preferences using a latent class approach, *Land Use Policy,* Vol. 68, 277-286.

Nunes P., P. Kumar, T. Dedeurwaerdere (2014): *Handbook on the Economics of Ecosystem Services and Biodiversity*, Edward Elgar, Cheltenham.

Petteri V., D. D'Amato, M. Forsius, P. Angelstam, C. Baessler, P. Balvanera, B. Boldgiv, P. Bourgeron, J. Dick, R. Kanka, S. Klotz, M. Maass, V. Melecis, P. Petrık, H. Shibata, J. Tang, J. Thompson and S. Zacharias (2013): Using long-term ecosystem service and biodiversity data to study the impacts and adaptation options in response to climate change: insights from the global ILTER sites network, *Current Opinion in Environmental Sustainability* 2013, 5:53–66.

Power, A. (2010): Ecosystem services and agriculture: Tradeoffs and synergies. *Philos. Trans. R. Soc. Lond. B Biol. Sci.* 365, 2959–2971.

Scholes R, B. Reyers, R. Biggs, M. Spierenburg and A. Duriappah (2013): Multi-scale and cross-scale assessments of social–ecological systems and their ecosystem services, *Current Opinion in Environmental Sustainability,* 5:16–25.

Tchipev N., St. Bratanova - Doncheva, K. Gocheva, M. Zhiyanski, M. Mondeshka, J. Yordanov, I. Apostolova, D. Sopotlieva, N. Velev, E. Rafailova, J. Uzunov, V. Karamfilov, Radka Fikova, St. Vergiev (2017): *Methodological framework for assessment and mapping of the state of ecosystems and ecosystem services in Bulgaria guide for monitoring the state and development of ecosystems and ecosystem services, EEA.*

Todorova K. (2017): Adoption of ecosystem-based measures in farmlands – new opportunities for flood risk management, *Trakia Journal of Sciences,* Vol. 15, 1, 152-157.

Todorova K. (2017): Flood risk management through ecosystem services from agricultural holdings, Dissertation, UNWE, WWF (2019): *Ecosystems and their "services," WWF.*

Tsiafouli M., E. Drakou, A. Orgiazzi, K. Hedlund and K. Ritz (2017): Optimizing the Delivery of Multiple Ecosystem Goods and Services in Agricultural Systems, *Frontiers in Ecology and Evolution*, vol. 5, art. 9715.

UN (2005). *The Millennium Development Goals Report.* United Nations, New York.

Van Oudenhoven, A. (2020): *Quantifying the effects of management on ecosystem services*, https://www.wur.nl/en/show/Quantifying-the-effects-of-management-on-ecosystem-services.htm.

Wang S., B. Fu, Y. Wei, C. Lyle (2013): Ecosystem services management: an integrated approach, *Current Opinion in Environmental Sustainability,* 5:11–15.

Wood S., D. Karp, F. DeClerck, C. Kremen, S. Naeem, C. Palm (2015): Functional traits in agriculture: agrobiodiversity and ecosystem services, *Trends in Ecology & Evolution,* 1–9.

Yordanov Ya., D. Mihalev, V. Vassilev, S. Bratanova-Doncheva, K. Gocheva, N. Chipev (2017): *Methodology for assessment and mapping of the state of agricultural ecosystems and their services in Bulgaria, EEA.*

Zhan J. (Editor) (2015): *Impacts of Land-use Change on Ecosystem Services*, Springer.

In: Environmental Management
Editor: Miguel Fischer
ISBN: 978-1-68507-019-9

Chapter 2

REEXAMINING COMPETITIVENESS OF BULGARIAN FARMS

Hrabrin Bachev* and Nina Koteva
Institute of Agricultural Economics, Sofia, Bulgaria

ABSTRACT

In an effort to fill the existing gap regarding the definition of the competitiveness of agricultural holdings and the ways to measure it, the present research applies a holistic approach in the assessment of the competitiveness of agricultural holdings in Bulgaria as a whole, as well as in terms of their different specialization. Despite its importance and the continuing debates on the topic, there is still no consensus on what the competitiveness of farms is; how to measure the competitiveness of different organizations in agriculture; what the absolute and comparative competitiveness of different types of farms is; which are the critical factors for increasing the competitiveness at the current stage of development, etc. The multi-criteria assessment found that although the level of competitiveness of Bulgarian farms is overall good, more than a third of all farms in the country show a low level of competitiveness. This can largely be attributed to their low adaptive potential and economic efficiency. The most competitive farms are those specializing in the beekeeping sector,

* Corresponding Author's E-mail: hbachev@yahoo.com.

followed by field crops, mixed livestock and mixed crop production, while farms specializing in grazing livestock are the least competitive. The proposed approach should be improved and applied more widely and periodically, increasing its accuracy and representativeness. The latter requires close cooperation with producer organizations, the National Agricultural Advisory Service (NAAS) and other stakeholders, as well as improvements to the agricultural information collection system in the country.

Keywords: competitiveness, agricultural holdings, Bulgaria

JEL: Q1; Q11; Q12; Q13; Q14; Q15; Q17; Q18

INTRODUCTION

The problem of determining the competitiveness of various economic organizations has been among the most topical academic and practical (aimed at improving business strategies and policies) issues since the emergence of economics science (Falciola and Rollo, 2020; Dresch et al., 2018; Westeren, et al., 2020; Wisenthige and Guoping, 2016). It is particularly important for the agricultural sector, which is characterized by many participants (including foreign ones), high specialization and exchange, strong competition at a local, national and international level, and highly integrated food and supply chains. Moreover, this sector has a number of specificities such as the dominance of small property ownership and informal management; the existence of quasi-monopoly situations in supply and sales; strong dependence on natural conditions; unequal public support; market segmentation; strong state regulation; processing and trade chains, professional organizations, etc.; strong consumer pressure for quality, eco-behavior, etc.; presence of underdeveloped and non-competitive "markets"; the need for new approaches, etc.

The problem of competitiveness has become particularly relevant in recent decades as a result of the fundamental development of the Theory of Economic Organizations (Bachev, 2012; Porter, 1980; Williamson, 1996), the processes of globalization and competition, and the new social and market "order" as defined by international agreements and institutions (World Trade Organization, World Bank,

International Monetary Fund, European Union, EC; FAO; OECD etc.). The latest processes such as the COVID-19 pandemic, climate change, the fundamental reform and "greening" of the Common Agricultural Policy (CAP) of the European Union (EU), widespread digitalisation, etc. pose new challenges to the competitiveness of agricultural producers in the country and around the world.

Despite its importance and the long-term lively discussions on the topic, there is still no consensus on: what is the competitiveness of agricultural holdings; how to measure the competitiveness of different organizations in agriculture; what is the absolute and comparative competitiveness of different types of agricultural farms; which are critical factors for increasing the competitiveness at the current stage of development, etc. Addressing all these issues is not just an important research matter, but a question of concern to farm managers and owners, professional and non-governmental organizations, politicians and the general public. It is no coincidence that increasing the viability and competitiveness of the sectors and agricultural producers has again been identified as one of the strategic objectives of the EU CAP for the new programming period 2021-2027 (EU, 2018).

Numerous studies have emerged in recent years on various aspects of the competitiveness of farms of different (mostly small) sizes (Alam et al., 2020; Berti and Mulligan, 2016; Latruffe, 2010, 2013; Lundy, et al., 2010; Mmari, 2015; Ngenoh et al., 2019; Orłowska, 2019), in selected countries (Alam et al., 2020; Benson, 2007; Jansik and Irz, 2015; Hadley, 2006; Popovic, Knezevic and Tosin, 2009; Kleinhanss, 2020; Krisciukaitiene, Melnikiene, Galnaityte, 2020; Nivievskyi, et al., 2011; Nowak, 2016; Mykhailova et al., 2018; Orłowska, 2019; Ziętara, Adamski, 2018), subsectors (Alam et al., 2020; Benson, 2007; FAO, 2010; Jansik and Irz, 2015; Kleinhanss, 2020; Marques et al., 2011; Marques, 2015; Nivievskyi, et al., 2011; Ngenoh et al., 2019; Oktariani, Daryanto, and Fahmi, 2016; Ziętara and Adamski, 2018), farming systems, such as organic, vertically integrated, greenhouse, etc. (Marques, 2015; Orłowska, 2019), regions (Marques et al., 2011; Nowak, 2016) and product chains (Lundy, et al., 2010; Ngenoh et al., 2019). A number of comparative studies have been conducted in different EU countries (FAO, 2010; Jansik and Irz, 2015; Nowak and Krukowski, 2019; Ziętara and Adamski, 2018) as a means of determining the technological, institutional and organizational factors for improving farm competitiveness (Berti, Mulligan, 2016; Mmari, 2015; Ngenoh et al., 2019; Oktariani, Daryanto, and Fahmi, 2016; OECD, 2011), etc.

To date, however, there is no widely accepted and comprehensive framework for understanding and assessing the competitiveness of farms in different market, economic, institutional and natural environments. Usually, the competitiveness of agricultural holdings is not well-defined and is assessed through traditional indicators of technical efficiency, productivity, profitability, etc. Rarely is a systematic approach applied to the formulation of pillars and the principles of competitiveness; to the criteria and indicators for evaluating its level; to the integration and interpretation of assessments, etc. Moreover, important aspects of farm competitiveness such as management efficiency, potential and incentives for adaptation, and "long-term" sustainability are often completely ignored in the analyses.

In Bulgaria, modern research on the absolute and comparative competitiveness of agricultural holdings is at the beginning stage (Andonov, 2013; Alexiev, 2012; Borisov, 2007; Bachev, 2010, 2010a, 2011, 2017; Ivanov et al., 2020; Koteva and Bachev, 2010, 2021; Koteva, 2016; Koteva et al., 2018; Slavova et al., 2011;). The number of publications on the level of competitiveness of agricultural holdings at the stage of EU CAP implementation is insignificant. In addition, there are practically no comprehensive studies on the competitiveness of farms with different product specializations at the current stage of development of the sector. This deters both the farms' management and the improvement of public support policies for the different types of producers. This article tries to fill in the existing gap by applying a holistic approach and assessing the competitiveness of farms in Bulgaria, both as a whole and in terms of their different specializations.

RESEARCH METHODOLOGY

Competitiveness means the internal capability (potential, incentives) of the agricultural holding to maintain sustainable competitive positions on (certain) market(s), which leads to high economic performance through continuous improvement and adaptation to the changing market, natural and institutional environment (Bachev, 2010; Koteva and Bachev, 2010, 2021). The level of competitiveness is always specific to a particular market-oriented farm in relation to the markets in which it sells its products and services.

Efficiency, financial endowment, adaptability and sustainability are the main pillars of the competitiveness of agricultural holdings. *Good* competitiveness means that a farm (1) produces and sells its products and services efficiently on the market; (2) manages its financing efficiently; (3) is adaptable to the evolving market, institutional and natural environment; and (4) is sustainable in time (Bachev, 2010; Koteva and Bachev, 2010). Conversely, insufficient (lack of) competitiveness indicates that the farm has serious problems in terms of efficient financing, the production and sale of its products due to high production and/or transaction costs, its inability to adapt to the evolving conditions of the environment and/or its insufficient sustainability over time.

For assessing the particular and integral level of competitiveness of Bulgarian farms, a holistic approach is applied, which includes a system of 4 criteria and 17 indicators and reference values, taking into account the economic efficiency, financial capabilities, adaptation potential and level of sustainability of farms. The choice of appropriate reference values is particularly important for an adequate assessment of the level of competitiveness. For example, a significant overpassing of the sectoral productivity and profitability is a sign of (higher) efficiency and competitiveness of farms; while a lack of "sufficient" liquidity shows low financial capability and low (lack of) competitiveness; the serious problems of the marketing of the produce and the lack of an heir willing to take over the farm are a sign of low sustainability and competitiveness, etc. A detailed presentation of the applied holistic approach, and the criteria for the selection and integration of the indicators for assessing the competitiveness of farms in Bulgaria is presented by Bachev (2010) and Koteva and Bachev (2010; 2021).

There is a lack of adequate (statistical and other) information in the country for assessing the various aspects of competitiveness of agricultural farms. In this study, the assessment of the level of competitiveness of farms is based on primary (survey) micro information provided in the summer of 2020 by the managers of 319 "typical" farms[1] of different types, production specializations and geographical locations. The structure of the surveyed farms approximately corresponds to the real structure of the farms in the country and in the main sub-sectors of the agricultural production sector in Bulgaria.

The farm managers were given the opportunity to indicate one of the three levels (low, good, high) which most closely corresponds to the condition of their holding for each indicator of the four competitiveness criteria. The qualitative assessments of the managers were transformed into quantitative values, as the high levels were assessed with 1, the intermediate with 0.5, and the low with 0. For each of the agricultural holdings, an integral competitiveness index was calculated for the individual criteria and as a whole as an arithmetic average. The competitiveness indices of farms with different types of specializations were obtained as an arithmetic average of the individual indices of the constituent holdings. To determine the overall level of competitiveness, the following benchmarks, set up by leading experts in the field, were used: high level 0.51-1, good level 0.34-0.5 and low level 0-0.32.

THE LEVEL OF COMPETITIVENESS OF BULGARIAN FARMS

The multi-criteria assessment of the competitiveness of agricultural holdings in the country shows that it is at a good level with a competitiveness index of 0.4 (Figure 1). The relatively high sustainability of farms (index 0.49) and, to a lesser extent, their good financial endowment (index 0.41) contribute the most to maintaining this level of competitiveness. On the other hand, the adaptability of agricultural holdings is relatively lower (index 0.39), and their economic efficiency is low (index 0.29). Therefore, the low potential for adaptation and the unsatisfactory economic efficiency contribute to the greatest extent to the decrease in the competitiveness of the Bulgarian farms, as they are critical for maintaining its level and restrict its increase.

---

[1] The authors thank the National Agricultural Advisory Service for their assistance and thank all the managers of the surveyed farms for the information provided.

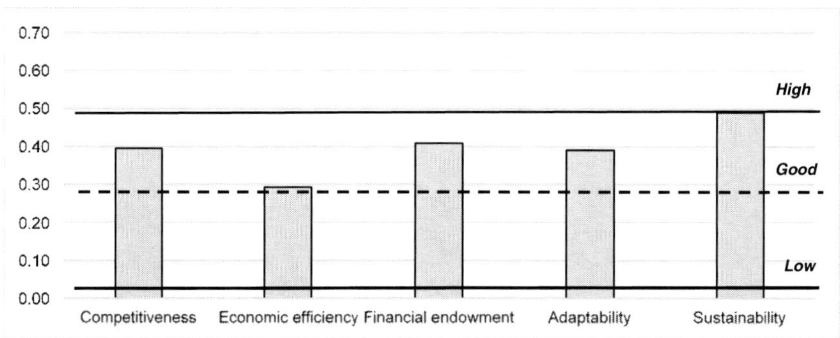

Source: Author's calculations.

Figure 1. Level of competitiveness of agricultural holdings in Bulgaria.

The analysis of the individual indicators of competitiveness shows the factors that most contribute to or limit the competitiveness of agricultural holdings in the country. At the present stage, the increase in the competitiveness of farms is limited by their extremely low productivity (0.16), profitability (0.19), financial capability (0.31) and adaptability to changes in the natural environment (rising temperatures, extreme weather, droughts, storms, etc.) – 0.33 (Figure 2). Therefore, both public support for farms and their management development strategies should be focused on these areas, as they are crucial for their competitiveness.

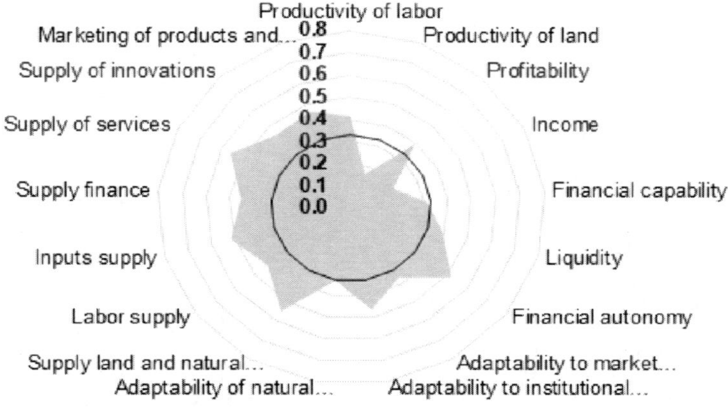

Source: Author's calculations.

Figure 2. Indicators of competitiveness of agricultural holdings in Bulgaria.

On the other hand, a number of indicators of the competitiveness of farms are at a high level and show the comparative and absolute competitive advantages of Bulgarian farms. At the present stage, the lack of serious problems and difficulties in the efficient supply of necessary services (0.56), the efficient supply of land and natural resources (0.55), the efficient supply of materials, equipment and biological resources (0.51), and the low dependence on external financing (credit, state aid, etc.) or high financial autonomy (0.52) contribute to increasing the competitiveness of agricultural holdings to the greatest extent.

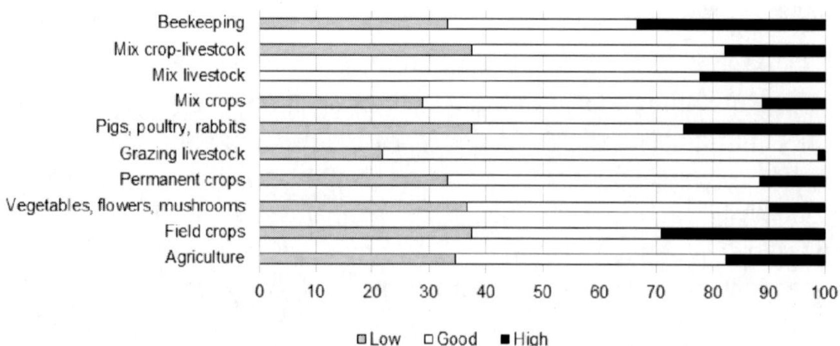

Figure 3. Share of agricultural holdings with different levels of competitiveness in Bulgaria (in %).

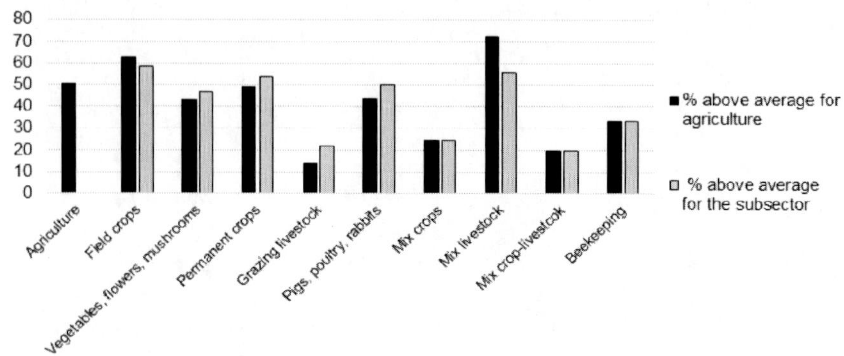

Source: Author's calculations.

Figure 4. Share of agricultural holdings with a level of competitiveness above the average for agriculture and the sub-sector in Bulgaria.

The assessment of the competitiveness of agricultural holdings shows that the majority of them (47.65%) have a good level of competitiveness (see Figure 3). Slightly more than half of the Bulgarian farms (50.47%) have a level of competitiveness above the national average (see Figure 4), and only 17.55% of all farms in the country have a high level of competitiveness. At the same time, however, more than a third of all farms (34.8%) have a low level of competitiveness. This means that a large part of Bulgarian farms will cease to exist in the near future due to insufficient competitiveness if timely measures are not taken to increase their competitiveness by improving their management and restructuring, through adequate state support, etc.

The vast majority of managers surveyed (64%) rated the competitiveness of their farms as good (Figure 5). The self-assessment of a large part of the managers differs from the multicriteria assessment made in this study, as the deviations are in both directions. Every tenth manager underestimates the (higher) level of competitiveness of their farm, and about 5% overestimate it. This means that independent multi-criteria assessments of competitiveness for the real situation would raise awareness and improve the management of a significant part of the farms in the country.

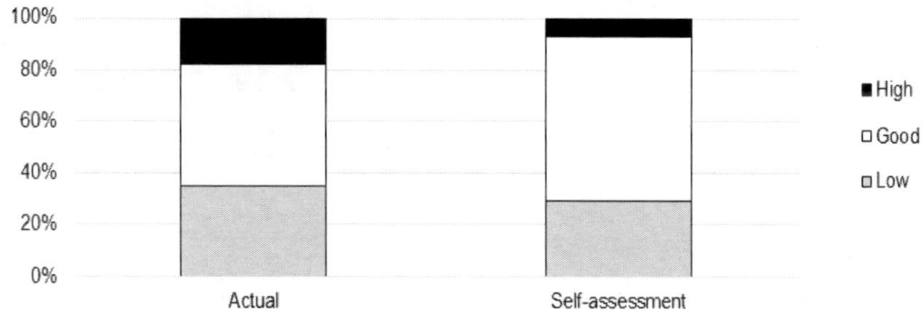

Source: Author's calculations; survey of agricultural producers, 2020.

Figure 5. Comparison of the multicriteria assessment with the self-assessment of the managers of the competitiveness of the agricultural holdings in Bulgaria.

The analysis of the share of farms with different levels of competitiveness indicators gives a clear idea of the situation in the country. The majority of Bulgarian farms have productivity and profitability, well below the national average – 68.54% and 62.79%, respectively (Table 1). Also, a significant part of the farms have low financial capability (38.02%), high dependence on external financing (loans, subsidies, etc.) (23.95%) and a low ability to pay their current liabilities (26.58%) (Table 2). In addition, 31.65% of the farms in the country have low adaptability to changes in the market environment (demand, prices, competition, etc.), 18.99% have insufficient adaptability to the institutional environment and constraints (national and European requirements for quality, safety, environment, etc.), and 36.39% have a low ability to adapt to changes in the natural environment (rising temperatures, extreme weather, drought, storms, etc.) (Table 3).

The survey also found that a significant part of the farms in the country have serious problems with the effective provision of the necessary labor force (30.5%), the necessary financing (20.89%), the necessary innovations and know-how (27.30%), and the effective marketing of production and services (18.85%) (Table 4). In addition, for every tenth farm there are major problems in the efficient supply of the necessary materials, equipment and biological resources (10.13%), for every ninth – in the effective supply of the necessary land and natural resources (8.68%), and for every seventh – in the effective supply of the necessary services (7.30%). All this contributes significantly to reducing the sustainability and competitiveness of a significant part of the holdings in the country.

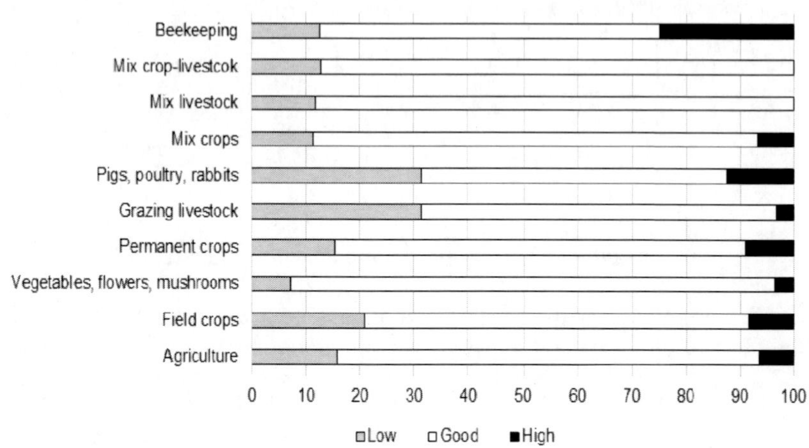

Source: Survey of agricultural producers, 2020.

Figure 6. How do you assess the sustainability of the agricultural holding in the medium term?

Table 1. Share of agricultural holdings with different levels of the indicators of economic efficiency in Bulgaria (in %)

Indicators levels	Agriculture	Field crops	Vegetables, flowers and mushrooms	Permanent crops	Grazing livestock	Pigs, poultry and rabbits	Mixed crops	Mixed livestock	Mixed crops and livestock	Beekeeping
Productivity										
Low	22.40	12.50	13.79	30.77	28.13	31.25	18.18	11.11	23.21	33.33
Good	71.92	70.83	82.76	61.54	71.88	62.50	81.82	83.33	75.00	44.44
High	5.68	16.67	3.45	7.69	0.00	6.25	0.00	5.56	1.79	22.22
Profitability										
Unsatisfactory	25.55	16.67	17.24	32.05	31.25	25.00	22.73	16.67	28.57	44.44
Good	69.40	70.83	79.31	61.54	68.75	75.00	75.00	77.78	69.64	33.33
High	5.05	12.50	3.45	6.41	0.00	0.00	2.27	5.56	1.79	22.22
*Gross output**										
Similar to the avarage	10.93	16.67	10.71	9.86	3.13	0.00	20.45	6.67	3.57	28.57
A little more than the avarage	3.64	12.50	3.57	4.23	3.13	0.00	0.00	0.00	5.36	0.00
A lot more than the avarage	1.32	0.00	0.00	1.41	0.00	0.00	2.27	0.00	3.57	0.00
A little less than the avarage	15.56	25.00	7.14	11.27	12.50	6.67	22.73	26.67	17.86	0.00
A lot less than the avarage	68.54	45.83	78.57	73.24	81.25	93.33	54.55	66.67	69.64	71.43

Table 1. (Continued)

Indicators levels	Agriculture	Field crops	Vegetables, flowers and mushrooms	Permanent crops	Grazing livestock	Pigs, poultry and rabbits	Mixed crops	Mixed livestock	Mixed crops and livestock	Beekeeping
*Net Income***										
Similar to the avarage	10.63	16.67	10.71	9.72	0.00	0.00	20.93	0.00	5.36	28.57
A little more than the avarage	4.65	12.50	3.57	6.94	3.23	0.00	0.00	6.67	5.36	0.00
A lot more than the avarage	1.66	0.00	0.00	2.78	0.00	0.00	2.33	0.00	3.57	0.00
A little less than the avarage	20.27	29.17	3.57	15.28	16.13	20.00	30.23	33.33	17.86	14.29
A lot less than the avarage	62.79	41.67	82.14	65.28	80.65	80.00	46.51	60.00	67.86	57.14

* Avarage Gross Output for the country = BGN 133,200; ** Avarage Net Income for the country = BGN 38,000.

Source: Survey of agricultural producers, 2020.

Table 2. Share of agricultural holdings with different levels of the indicators for financial endowment in Bulgaria (in %)

Indicators levels	Agriculture	Field crops	Vegetables, flowers and mushrooms	Permanent crops	Grazing livestock	Pigs, poultry and rabbits	Mixed crops	Mixed livestock	Mixed crops and livestcok	Beekeeping
Financial capability										
Low	38.02	26.09	46.43	40.26	51.61	50.00	28.89	22.22	39.29	44.44
Good	61.34	73.91	53.57	59.74	48.39	50.00	71.11	77.78	58.93	44.44
High	0.64	0.00	0.00	0.00	0.00	0.00	0.00	0.00	1.79	11.11
Dependance on external financing (credit, state support, etc.)										
Low	27.83	30.43	28.57	28.38	28.13	26.67	25.58	16.67	30.36	33.33
Avarage	48.22	52.17	46.43	50.00	40.63	46.67	46.51	55.56	44.64	55.56
High	23.95	17.39	25.00	21.62	31.25	26.67	27.91	27.78	25.00	11.11
Possibility to pay current debts										
Low	26.58	25.00	31.03	24.68	43.75	33.33	15.56	22.22	32.14	22.22
Good	68.04	66.67	65.52	71.43	56.25	66.67	73.33	72.22	66.07	55.56
High	5.38	8.33	3.45	3.90	0.00	0.00	11.11	5.56	1.79	22.22

Table 3. Share of agricultural holdings with different levels of the indicators of adaptability (in %)

Indicators levels	Agriculture	Field crops	Vegetables, flowers and mushrooms	Permanent crops	Grazing livestock	Pigs, poultry and rabbits	Mixed crops	Mixed livestock	Mixed crops and livestcok	Beekeeping
Adaptability to the market (prices, demand, competition)										
Low	31.65	25.00	17.24	37.66	50.00	25.00	24.44	33.33	33.93	33.33
Good	62.66	62.50	72.41	59.74	46.88	62.50	73.33	61.11	64.29	33.33
High	5.70	8.33	10.34	3.90	3.13	12.50	2.22	5.56	0.00	33.33
Adaptability to the state and European requirements for quality, safety, environment, etc.										
Low	18.99	20.83	20.69	11.69	34.38	18.75	20.00	16.67	23.21	0.00
Good	68.35	66.67	72.41	77.92	65.63	62.50	64.44	50.00	66.07	66.67
High	12.66	12.50	6.90	10.39	0.00	18.75	15.56	33.33	8.93	33.33
Adaptability to changes in the natural environment (rising temperatures, extreme weather, drought, storms, etc.)										
Low	36.39	29.17	34.48	41.56	34.38	37.50	33.33	22.22	46.43	22.22
Good	60.44	66.67	65.52	55.84	59.38	62.50	64.44	61.11	51.79	66.67
High	36.39	29.17	34.48	41.56	34.38	37.50	33.33	22.22	46.43	22.22

Source: Survey of agricultural producers, 2020.

Table 4. Share of agricultural holdings with different levels of indicators of sustainability in Bulgaria (in %)

Indicators levels	Agriculture	Field crops	Vegetables, flowers and mushrooms	Permanent crops	Grazing livestock	Pigs, poultry and rabbits	Mixed crops	Mixed livestock	Mixed crops and livestcok	Beekeeping
Nature of the problems in the effective supply of necessary land and natural resources										
Insignificant	18.65	20.83	22.22	14.29	18.75	40.00	20.45	11.11	14.55	50.00
Normal	72.67	75.00	77.78	75.32	62.50	53.33	72.73	72.22	78.18	37.50
Significant	8.68	4.17	0.00	10.39	18.75	6.67	6.82	16.67	7.27	12.50
Nature of the problems in the effective supply of necessary labor force										
Insignificant	16.67	16.67	27.59	10.26	18.75	18.75	8.89	5.56	25.00	44.44
Normal	52.83	66.67	51.72	53.85	40.63	68.75	53.33	50.00	50.00	33.33
Significant	30.50	16.67	20.69	35.90	40.63	12.50	37.78	44.44	25.00	22.22
Nature of the problems in the effective supply of necessary materials, equipment and biological resources										
Insignificant	12.97	12.50	24.14	10.53	9.38	6.25	13.33	11.11	12.50	33.33
Normal	76.90	79.17	65.52	75.00	78.13	81.25	82.22	77.78	76.79	66.67
Significant	10.13	8.33	10.34	14.47	12.50	12.50	4.44	11.11	10.71	0.00
Nature of the problems in the effective supply of necessary funding										
Insignificant	12.03	4.17	10.34	15.58	9.68	0.00	13.33	16.67	14.29	22.22
Normal	67.09	83.33	58.62	70.13	54.84	87.50	57.78	72.22	62.50	77.78
Significant	20.89	12.50	31.03	14.29	35.48	12.50	28.89	11.11	23.21	0.00
Nature of the problems in the effective supply of necessary services										
Insignificant	18.41	8.33	27.59	21.05	15.63	25.00	15.56	16.67	19.64	22.22
Normal	74.29	79.17	72.41	71.05	75.00	62.50	80.00	72.22	73.21	77.78
Significant	7.30	12.50	0.00	7.89	9.38	12.50	4.44	11.11	7.14	0.00

Table 4. (Continued)

Indicators levels	Agriculture	Field crops	Vegetables, flowers and mushrooms	Permanent crops	Grazing livestock	Pigs, poultry and rabbits	Mixed crops	Mixed livestock	Mixed crops and livestcok	Beekeeping
Nature of the problems in the effective supply of necessary innovations and know-how										
Insignificant	17.46	16.67	14.29	21.79	18.75	18.75	17.78	23.53	12.50	11.11
Normal	55.24	58.33	57.14	61.54	37.50	50.00	53.33	52.94	55.36	88.89
Significant	27.30	25.00	28.57	16.67	43.75	31.25	28.89	23.53	32.14	0.00
Nature of the problems in the effective realization of the products and services										
Insignificant	12.46	20.83	17.86	14.29	6.45	12.50	11.11	5.56	10.71	12.50
Normal	68.69	66.67	71.43	63.64	67.74	62.50	75.56	83.33	67.86	62.50
Significant	18.85	12.50	10.71	22.08	25.81	25.00	13.33	11.11	21.43	25.00

Source: Survey of agricultural producers, 2020.

According to the managers of a large part of the farms in the country (15.71%), their farms have low sustainability in the medium term and are likely to cease to exist due to bankruptcy, cessation of business, acquisition by competitors, etc. (Figure 6).

The vast majority of managers (77.88%) evaluate the sustainability of their farms as good (Figure 7). In contrast to competitiveness, in the self-assessments of sustainability, there is almost a coincidence of the share of farms with low sustainability with that of the multi-criteria assessment in the study. However, there is a significant underestimation of the level of "real" sustainability in the self-assessment of the managers of farms with high sustainability – a little over 5 times. This means that many farm managers do not have an accurate idea of the real level of (economic) sustainability of the farms they manage. Therefore, holistic "external" sustainability assessments, such as the one in this study, would greatly improve the awareness, self-confidence and overall management of a significant part of the country's farms.

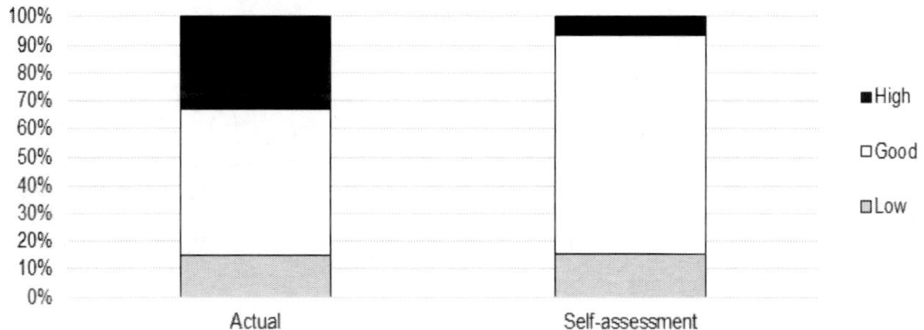

Source: Author's calculations; survey of agricultural producers, 2020.

Figure 7. Comparison of the multicriteria assessment with the self-assessment of the managers of the sustainability of the agricultural holdings in Bulgaria.

Level of Competitiveness of Farms with Different Specializations

There is a significant variation in the level of competitiveness of agricultural holdings with different production specializations (Figure 8).

The farms with the highest *good level of* competitiveness are in the beekeeping sector (0.46), followed by those specialed in field crops (0.44), mixed livestock (0.42), and mixed crop production (0.41). The farms in a number of major agricultural sub-sectors are with a good level of competitiveness, however, it is below the national average – permanent crops (0.39), vegetables, flowers and mushrooms (0.38), pigs, poultry and rabbits (0.38), and mixed crops and livestock (0.38). The farms specializing in grazing livestock are the least competitive with a *low* level (0.32) of competitiveness.

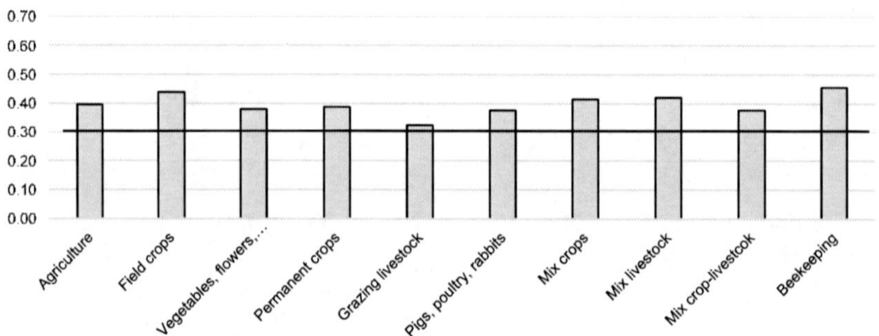

Source: Author's calculations.

Figure 8. Competitiveness of agricultural holdings with different specializations in Bulgaria.

The analysis of the individual aspects of the competitiveness of farms with different specializations shows that most types have low economic efficiency, which is the largest contributor to the deterioration of their competitiveness (Figure 9). Only the farms specializing in field crops have good economic efficiency.

The farms specialized in beekeeping (0.48) have the best financial endowment, followed by field crop (0.45) and mixed crop farms (0.44). The financial endowment of farms specialized in mixed crops and livestock production (0.4), vegetables, flowers and mushrooms (0.38), pigs, poultry and rabbits (0.36) and grazing livestock (0.34) is below the national average, the latter group being close to the low level.

The farms specialized in beekeeping (0.54), mixed livestock (0.47) and pigs, poultry and rabbits (0.42) have the highest adaptability. The potential for adaptation to changes in the market, institutional and natural environment of farms specializing in permanent crops (0.38) and mixed crops and livestock (0.35) is below the industry average, while that of farms with grazing livestock is at a low level (0.3).

The sustainability of most types of farms is relatively good and close to the national average. The farms with the lowest sustainability within the limits of the good level are those specialized in grazing livestock (0.44). The sustainability of the other groups of farms is at a high level, with a maximum value for those specialized in beekeeping.

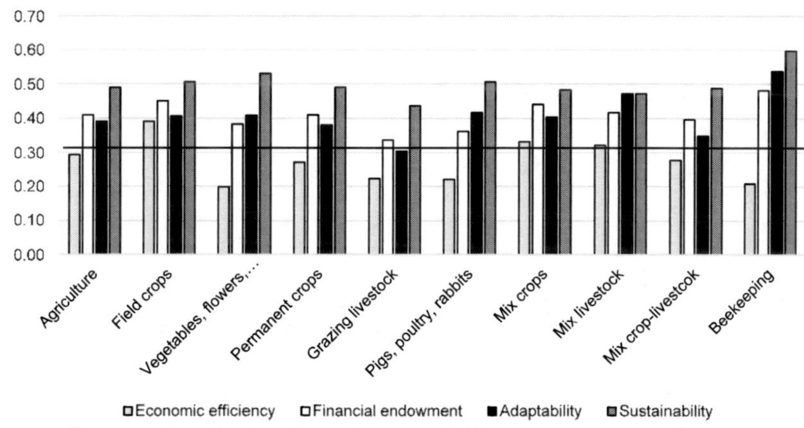

Source: Survey of agricultural producers, 2020.

Figure 9. Level of competitiveness of agricultural holdings with different specializations by main criteria for competitiveness in Bulgaria.

Most of the indicators of competitiveness of farms specializing in *field crops* have values higher than the national average (Figure 10).These farms have lower than average levels only in terms of their adaptability to the institutional environment and the efficiency of service provision.

The competitiveness of farms specializing in the cultivation of field crops is maintained by high productivity, liquidity, financial autonomy, adaptability to the market environment, efficiency in the supply of land and natural resources, materials, machinery and biological resources, finance, services and innovation, and efficient realization of products and services. The main factors for reducing the competitiveness of farms with field crops are low productivity (0.27) and profitability (0.29), as well as adaptability to the natural environment (0.35) that is close to the low level.

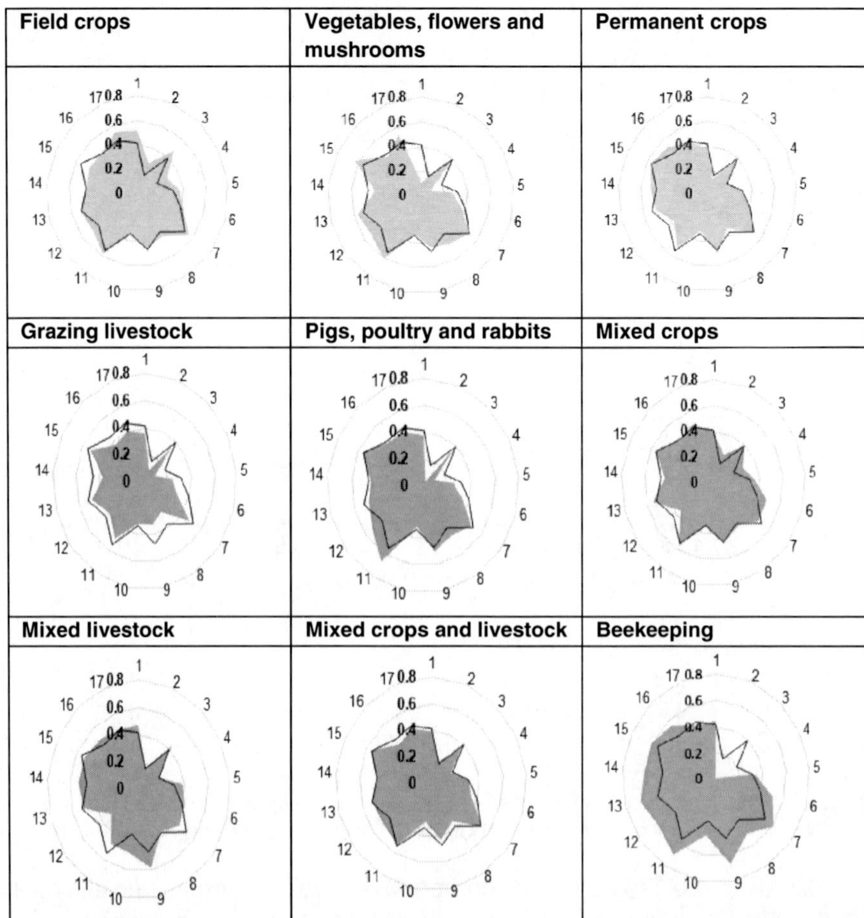

* 1 – Productivity of labor; 2 – Productivity of land; 3 – Profitability; 4 – Income; 5 – Financial capability; 6 – Liquidity; 7 – Financial autonomy; 8 – Adaptability to the market environment; 9 – Adaptability to the institutional environment; 10 – Adaptability to the natural environment; 11 – Supply of land and natural resources; 12 – Labor supply; 13 – Inputs supply; 14 – Financial supply; 15 – Supply of services; 16 – Supply of innovations; 17 – Marketing of products and services.

Source: Author's calculations.

Figure 10. Level of the indicators* for competitiveness of agricultural holdings with different specialisations in Bulgaria (the avarage for agriculture is presented in black).

Many of the indicators of the competitiveness of farms specializing in the cultivation of vegetables, flowers and mushrooms have values lower than the national average (Figure 10). However, in many respects, these farms have higher than average positions – in terms of their profitability, adaptability to the market environment, efficiency in the supply of land and natural resources, labor, materials, machinery and biological resources, services, and the sale of products and services. The main factors for maintaining the competitive position of this type of farm are high financial autonomy, efficiency in the supply of land and natural resources, labor, materials, equipment and biological resources, services and the sale of products and services. The main factors for reducing their competitiveness are low labour productivity (0.11), land productivity (0.16), profitability (0.09), financial capability (0.27) and adaptability to the natural environment (0.33).

The majority of indicators for the competitiveness of farms specialized in the cultivation of *permanent crops* have values lower than the national average (Figure 10). However, in some areas, these farms have better-than-average positions, such as in terms of their financial autonomy, adaptability to the institutional environment and their efficiency in the supply of finance, services and innovation. The competitiveness of this type of farm is maintained by high financial autonomy, adaptability to the institutional environment, efficiency in the supply of land and natural resources, services and innovation. The most important factors for the deterioration of the competitive position of the farms specializing in the cultivation of permanent crops are low productivity (0.14), profitability (0.19), financial capability (0.3), adaptability to the market environment (0.33) and the natural environment (0.31).

All indicators of the competitiveness of farms specializing in grazing livestock have values lower than the national average (Figure 10). The low productivity (0.09), profitability (0.1), financial capability (0.24), liquidity (0.28) and adaptability to the market (0.27), institutional (0.33) and natural (0.32) environment contribute the most to the unsatisfactory competitiveness of this type of farm. The main factor for raising their competitive position is the high efficiency in their supply of services.

Most of the competitiveness indicators of farms specializing in pigs, poultry and rabbits have values lower than the national average (Figure 10). However, in several respects, these farms have better-than-average positions, such as in terms of their adaptability to the market and institutional environment, and their efficiency in the supply of land and natural resources, labor and services. The most important factors for maintaining the competitiveness of this type of farm are the high efficiency in the supply of land and natural resources, labor and services. Low productivity (0.03), profitability (0.1), financial capability (0.25), liquidity (0.33) and adaptability to changes in the natural environment (0.31) are critical for maintaining the competitive positions of farms specializing in pigs, poultry and rabbits.

Many of the indicators of the competitiveness of farms specializing in mixed crop production have values lower than the national average (Figure 10). However, in many areas, this type of farm has relatively better-than-average positions, such as in terms of its profitability, financial capability, liquidity, adaptability to the market, institutional and natural environment, and efficiency in the supply of land and natural resources, materials, equipment and biological resources, as well as in the realization of products and services. Central to maintaining the competitiveness of these farms are their high efficiency in the supply of land and natural resources, materials, machinery and biological resources and services. At the same time, however, the competitive position of mixed crop farms is compromised by low productivity (0.24) and income (0.28), and a close-to-low level of adaptability to changes in the natural environment (0.34).

Many of the competitiveness indicators of mixed livestock farms are higher than the national average (Figure 10). The farms specialized in this field are superior to other farms in terms of their productivity, profitability, financial capability, liquidity, adaptability to the institutional and natural environment, efficiency in the supply of finance and innovation, as well as in the sale of products and services. The other indicators of competitiveness of this type of farm are lower or around the average levels for the country. The high adaptability to the institutional environment and the efficiency in the supply of finances and services contribute the most to maintaining the competitive positions of the mixed livestock farms. At the same time, however, the indicators of productivity (0.17), profitability (0.2) and efficiency in labor supply (0.31) are low and limit the improvement of the overall competitiveness of these farms.

Almost all indicators of the competitiveness of mixed crop and livestock farms are lower or close to the national average (Figure 10). These farms are above average only in terms of their financial autonomy and their efficiency in the supply of labor and services. High financial autonomy and efficiency in the supply of land and natural resources, materials, machinery and biological resources and services contribute the most to maintaining the competitive position of this type of farm. At the same time, low productivity (0.17), profitability (0.18), financial capability (0.31), and adaptability to changes in the market environment (0.33) and the natural environment (0.29) are critical for the competitiveness of mixed crop and livestock farms.

Almost all indicators of the competitiveness of farms specializing in *beekeeping* are higher than the national average, with the exception of the indicators of productivity, profitability, income and efficiency in the sale of products and services (Figure 10). The competitiveness of this type of farm benefits from its high level of financial autonomy, adaptability to the institutional environment, efficiency in the supply of resources, services and innovation. At the same time, however, low productivity and profitability are the factors that worsen the competitive position of beekeepers.

The assessment of the competitiveness for agricultural holdings shows that the majority of those specialized in *field crops* (62.5%) and *mixed livestock* (72.22%) have a level of competitiveness above the national average (Figure 4). The lowest share of farms with competitiveness exceeding the national average is in the sectors of *grazing livestock* (14.1%), mixed crops and livestock (19.64%), mixed crops (24.44%) and beekeeping (one third).

There are also big differences in the share of farms with different types of specializations which exceed the average level of competitiveness for their respective sub-sector (type). While farms specialized in field crops 58.33% are competitive above the average for this sector, in the case of mixed crop and livestock farms they are only 19.64% (Figure 4). The share of farms with a level of competitiveness superior to that of the grazing livestock sector (21.79%) and the beekeeping sector (one third) is also very low.

The largest share of farms with high competitiveness is in the sectors of beekeeping (one third), field crops (29.17%), pigs, poultry and rabbits (a quarter) and mixed livestock (22.22%); and the smallest share is occupied by farms specialized in grazing livestock – only 1.28% (Figure 3). At the same time, the share of farms with low competitiveness in each type of specialization is significant – field crops, pigs, poultry and rabbits, and mixed crops and livestock make up 37.5% each; vegetables, flowers and mushrooms make up 36.67%, permanent crops and bees make up 33.33%, mixed crops make up 28.89%, and grazing livestock – 21.79%. Only in the group of mixed livestock farms there are no ones with low competitiveness.

There is a discrepancy between the assessments of the level of competitiveness in the present analysis and the self-assessments of the managers of the surveyed farms with different specializations (Figure 11). While the majority of beekeepers (37.50%) believe that their farms are highly competitive, in other groups of farms this percentage is much lower – from 1.8% (mixed crops and livestock) to 9% (permanent crops). None of the managers operating in the field crops sector puts their farm in the group of the highly competitive ones. At the same time, the share of managers who assess their farm as having a low level of competitiveness is large – 30.43% for field crops, 21.43% for vegetables, flowers and mushrooms, 28.21% for permanent crops, 46.88% for grazing livestock, 31.25% for pigs, poultry and rabbits, 22.22% for mixed crops, 27.78% for mixed livestock, 35.71% for mixed crops and livestock, and 12.5% for beekeeping. Therefore, independent multi-criteria evaluations such as those in the present study would improve the awareness and management of farms that overestimate or underestimate their actual competitiveness.

The survey of managers found that there are large differences in the share of farms with each type of specialization with different levels of competitiveness indicators. A significant part of the farms in all subsectors have productivity and profitability well below the national average (Table 3). In addition, a large proportion of farms specializing in permanent crops, pigs, poultry and rabbits, and beekeeping have low productivity and profitability.

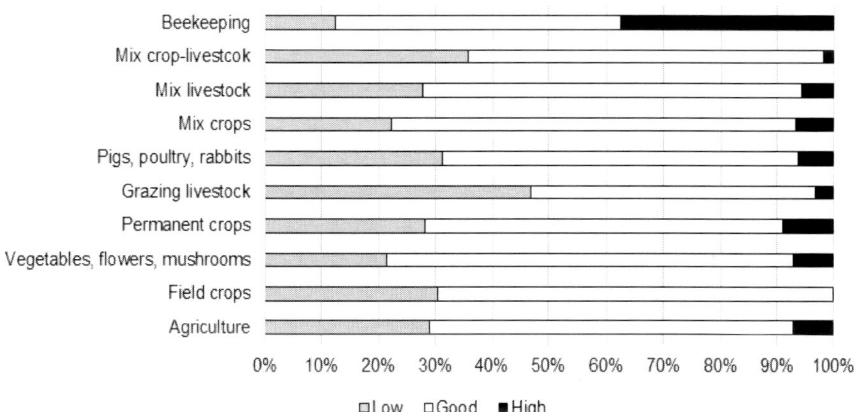

Source: Survey of agricultural producers, 2020.

Figure 11. How do you assess the competitiveness of the agricultural holding?

The largest share of farms with low financial capability is in the following sectors: vegetables, flowers and mushrooms (46.43%), permanent crops (40.26%), grazing livestock (51.61%), pigs, poultry and rabbits (50%), and beekeeping (44.44%) (Table 4). Most farms with high dependence on external financing (loans, subsidies, etc.) are in the groups of grazing livestock (31.25%), mixed crops (27.91%) and mixed livestock (27.78%). The most significant is the share of farms with low ability to pay their current obligations, which fall in the sectors: vegetables, flowers and mushrooms (31.03%), grazing livestock (43.75%), pigs, poultry and rabbits (every third) and mixed crops and livestock (32.14%).

Many farms with different types of specializations have insufficient potential to adapt to changes in the market, institutional and natural environment (Table 5). The largest share of farms with low adaptability to changes in the market environment (demand, prices, competition, etc.) are in the following sectors: permanent crops (37.66%), grazing livestock (every second), mixed livestock, mixed crops and livestock, and beekeeping (one third each). Most farms with insufficient adaptability to the institutional environment and restrictions (state and European requirements for quality, safety, environment, etc.) are among those specializing in grazing livestock (34.38%), and mixed crop and livestock farms (23.21%). There is also a significant share of farms with low ability to adapt to changes in the natural environment (rising temperatures, extreme weather, drought, sleet, etc.), which varies from 22.22% in the case of mixed livestock and bees, to 46.43% in the case of all mixed crop and livestock farms in the country.

The survey found that the largest share of farm managers who believe that their farms have low sustainability in the medium term are among those specializing in field crops (20.83%), grazing livestock, and pigs, poultry and rabbits (31.25%) (Figure 6).

The survey also found that a significant proportion of farms specializing in permanent crops (35.9%), grazing livestock (40.63%), mixed crops (37.78%) and mixed livestock (44.44%) have serious problems and difficulties in effectively providing the needed labor force (Table 6). There are also many farms that face serious problems and difficulties in effectively providing the necessary funding – 31.03% of all farms specialized in growing vegetables, flowers and mushrooms, 35.48% of those specialized in grazing livestock, and 28.89% of those specialized in mixed crops. In addition, a large part of farms specializing in grazing livestock (43.75%), pigs, poultry and rabbits (31.25%), and mixed crops and livestock (32.14%) are faced with serious problems and difficulties in effectively providing the necessary innovations and know-how. There are also many farms specializing in permanent crops (22.08%), grazing livestock (25.81%), pigs, poultry and rabbits, and beekeeping (a quarter each), which face serious problems and difficulties in the effective sale of their products and services.

Conclusion

The multi-criteria assessment of the level of competitiveness of agricultural holdings in Bulgaria found that it stands at a good level, as the low adaptive potential and economic efficiency contribute to the greatest extent to diminishing the competitiveness of local producers. Particularly critical for maintaining the competitive position of farms are the low productivity, profitability, financial capability and adaptability to changes in the natural environment, and those are the areas in which public support for farms and farm management development strategies should be directed. More than a third of all farms in the country have a low level of competitiveness, and if timely measures are not taken to increase their competitiveness by improving their management and restructuring them by providing adequate state support, etc., a large part of Bulgarian farms will cease to exist in the near future. The most competitive are the farms in the beekeeping sector, followed by those specialized in field crops, mixed livestock and mixed crop production, and the least competitive are the farms specialized in grazing livestock. The proposed approach to assessing the competitiveness of farms should be refined and applied more widely and periodically. The analyses should also cover holdings of different juridicial type, size, ecological and geographical location, etc. The accuracy and representativeness of the information used should also be enhanced by increasing the number of surveyed farms, applying statistical methods, special "training" of those conducting and participating in the surveys, etc. All this requires closer cooperation with producer organizations, the National Agricultural Advisory Service (NAAS) and other stakeholders, as well as improvements in the system for collecting agricultural information in the country.

REFERENCES

Alam, S., Munizu, M., Munir, A. R., Pono, M., Kadir, A. R. O. (2020). Development Model of Competitiveness of Chicken Farm SMEs in Sidrap Regency, South Sulawesi, Indonesia. *ESPACIOS*, Vol. 41 (Issue 10), 23.

Aleksiev, A. (2012). *The competitive opportunities of the grain sector.* Plovdiv: Academic PH of the Agricultural University (*in Bulgarian*).

Andonov, S. (2013). *The role of European subsidies in increasing the competitiveness of the agricultural sector in Bulgaria* (a doctoral thesis). Sofia University "St. Kliment Ohridski" (*in Bulgarian*).

Andrew, D., Semanik, M., Torsekar, M. (2018). *Framework for Analyzing the Competitiveness of Advanced Technology Manufacturing Firms.* Office of Industries Working Paper ID-057, September.

Atristain Suarez, C. (2013). *Organizational Performance and Competitiveness: Analysis of Small Firms*. Nova science Publisher.

Bachev, H. (2010). *Management of Farm Contracts and Competitiveness*. VDM Verlag Dr. Muller, Germany and USA.

Bachev, H. (2010a). Assessing Farm Competitiveness in Bulgaria. *Agriculture Economics and Management*, N 6, pp. 11-26 (*in Bulgarian*).

Bachev, H. (2011). Assessing the competitiveness of agricultural cooperatives. *Agriculture Economics and Management*, N 1, pp. 22-30 (*in Bulgarian*).

Bachev, H. (2011). Competitiveness of the agricultural holdings of natural persons. *Agriculture Economics and Management*, N 5, pp. 55-65 (*in Bulgarian*).

Bachev, H. (2012). The Efficiency of Farms and Agrarian Organizations. *Economic Thought Journal*, N 4, pp. 46-77 (*in Bulgarian*).

Bachev, H. (2017). The sustainability of management structures in Bulgarian agriculture – level, factors, prospects. *Economics 21*, pp. 69-95 (*in Bulgarian*).

Benson, G. (2007). *Competitiveness of NC Dairy Farms.* North Carolina State University, http://www.ag-econ.ncsu.edu/faculty/benson/DFPPNatComp01.PDF

Chursin, A. & Makarov, Y. (eds.). (2015). *Management of Competitiveness: Theory and Practice*. London: Springer.

Csaba, J., Irz, X. (2015). Competitiveness of Dairy Farms in Northern Europe: A Cross-Country analysis. *Agricultural and Food Science*, 24, 3, pp. 206-218.

Dresch, A, Collatto, D. C., Lacerda, (2018). Theoretical understanding between competitiveness and productivity: firm level. *Ingeniería y competitividad*, Vol.20, N 2, pp. 69-86. [*Engineering and competitiveness*]

EC (2018). *Proposal for a Regulation of the European Parliament and of the Council establishing rules on support for strategic plans to be drawn up by Member States under the Common agricultural policy* (CAP Strategic Plans) and financed by the European Agricultural Guarantee Fund (EAGF) and by the European Agricultural Fund for Rural Development (EAFRD) and repealing Regulation (EU) No. 1305/2013 of the European Parliament and of the Council and

Regulation (EU) No. 1307/2013 of the European Parliament and of the Council, European Commission, Brussels.

FAO (2010). *International Competitiveness of 'Typical' Dairy Farms.*

Falciola, J., Jansen, M., Rollo, V. (2020). Defining firm competitiveness: A multidimensional framework. *World Development, Elsevier,* Vol. 129(C).

Giaime, B. & Mulligan, C. (2016). Competitiveness of Small Farms and Innovative Food Supply Chains: The Role of Food Hubs in Creating Sustainable Regional and Local Food Systems. *Sustainability*, 8, 616.

Ivanov, B., Popov, R., Bachev, H., Koteva, N., Malamova, N., Chopeva, M., Todorova, K., Nacheva, I., Mitova, D. (2020). *Analysis of the state of agriculture and the food industry.* Sofia: Institute of Agricultural Economics (*in Bulgarian*).

Kleinhanss, W. (2020). Competitiveness of the Main Farming Types in Germany. In: *20th International Farm Management Congress*, Vol.1. IFMA.

Koteva, N., Bachev, H. (2010). Approach for assessing the competitiveness of agricultural holdings. *Agriculture Economics and Management,* N 1, pp. 32-43 (*in Bulgarian*).

Koteva, N., Bachev, H. (2011). A study of the competitiveness of agricultural holdings in Bulgaria. *Economic Thought Journal*, N 5, pp. 34-63 (*in Bulgarian*).

Koteva, N. (2016). *Development and competitiveness of agricultural holdings in Bulgaria in the conditions of the EU CAP*. Sofia: Avangard Prima (*in Bulgarian*).

Koteva, N., Aleksiev, A., Beluhova-Uzunova, R., Roicheva, A., Hadzichoneva, J., Georgiev, A., Hadjiev, K. (2018). Theoretical and Methodological Aspects of the Competitiveness of Agricultural Holdings. *Agriculture Economics and Management*, 63, 4, pp. 3-14 (*in Bulgarian*).

Koteva, N., Bachev, H. (2021). *The competitiveness of agricultural holdings in Bulgaria and models for its increase*. Sofia: Institute of Agricultural Economics (*in Bulgarian*).

Krisciukaitiene, I., R. Melnikiene, A. Galnaityte (2020): *Competitiveness of Lithuanian farms and their agriculture production from present to medium - term perspectives*, Lithuanian IAE.

Latruffe, L. (2010). Competitiveness, Productivity and Efficiency in the Agricultural and Agri-Food Sectors. OECD Food, Agriculture and Fisheries Papers, N 30. OECD Publishing.

Latruffe, L. (2013). Competitiveness in the agricultural sector: measures and determinants. *Farm Policy Journal*, 11(3), pp. 9-17.

Lundy, M., Gottret, M. V., Cifuentes, W., Ostertag, C. F., Best, R., Peters D., and Ferris, Sh. (2010). *Increasing the Competitiveness of Market chains for Smallholder producers*. CIAT.

Marques, P. R. et al. (2015): Competitiveness levels in cattle herd farms. *Ciencia Rural*, Vol.45, N.3, pp. 480-484.

Marques, P. R., Barcellos, J.O.J., McManus, C., Oaigen, R. P., Collares, F.C., Canozzi, M. E. A., Lampert, V. N. (2011). Competitiveness of beef farming in Rio Grande do Sul State, Brazil. *Agricultural Systems*, Vol. 104, Issue 9, pp. 689-693.

Mmari, D. (2015). *Institutional Innovations and Competitiveness of Smallholders in Tanzania*. Thesis to obtain the degree of Doctor from the Erasmus University Rotterdam.

Ngenoh, E., Kurgat, B. K., Bett, H., Kebede, S. W. and Bokelmann, W. (2019). Determinants of the competitiveness of smallholder African indigenous vegetable farmers in high-value agro-food chains in Kenya: A multivariate probit regression analysis. *Agricultural and Food Economics*, 7, pp. 2-17.

Nivievskyi, O., von Cramon-Taubadel, S. (2010). *The Determinants of Dairy Farming Competitiveness in Ukraine*. Policy Paper Series [AgPP No 23]. Institute for Economic Research and Policy Consulting.

Nowak A. (2016). Regional Differences in the Competitiveness of Farms in Poland. *Journal of Agribusiness and Rural Development*, 3, 41, pp. 345-354.

Nowak, A., Krukowski, A. (2019). Competitiveness of farms in new European Union member states. *Agronomy Science*, 2, pp. 73-80.

OECD (2011). *Fostering Productivity and Competitiveness in Agriculture*.

Oktariani, A., Daryanto, A., Fahmi, I. (2016): The competitiveness of dairy farmers based fresh milk marketing on agro-tourism. *International Journal of Animal Health and Livestock Production Research*, Vol.2, N 1, pp. 18-38.

Orłowska, M. (2019): Competitiveness of Pollish Organic Farms with Different Economic Size in Light of Fadn Data. *Annals PAAAE 2019*, XXI (2), pp. 217-224.

Porter, M. (1980). *Competitive Strategy: Techniques for Analyzing Industries and Competitors*. The Free Press, Macmillan.

Slavova, Y. et al. (2011). *Competitive opportunities of the agricultural sector*. Sofia: Agricultural Academy, Institute of Agricultural Economics (*in Bulgarian*).

Westeren, K. I., Cader, H., Sales, M. F., Similä, J. O., Staduto, J. (2020). *Competitiveness and Knowledge. An International Comparison of Traditional Firms*. Routledge.

Wisenthige, K., & Guoping, C. (2016). Firm level competitiveness of small and medium enterprises (SMEs): analytical framework based on pillars of competitiveness model. *International Research Journal of Management, IT and Social Sciences*, 3(9), pp. 61-67.

Williamson, O. (1996). *The Mechanisms of Governance*. New York: Oxford University Press.

Ziętara W., Adamski, M. (2018). Competitiveness of the Polish dairy farms at the background of farms from selected European Union countries, *Problems of Agricultural Economics*, 1(354), pp. 56-78.

In: Environmental Management
Editor: Miguel Fischer
ISBN: 978-1-68507-019-9

Chapter 3

Distribution and Composition of Metallic Mining Solid Wastes in the Central Mexico Region, and the Need for Adequate Environmental Management

Alexis Joavany Rodríguez-Solís[1], Natalia de la Cruz-Guarneros[2], María Luisa Castrejón-Godínez[3], Efraín Tovar-Sánchez[4] and Patricia Mussali-Galante[1,*]

[1]Centro de Investigación en Biotecnología, Universidad Autónoma del Estado de Morelos. Cuernavaca, Morelos, México
[2]Especialidad en Gestión Integral de Residuos, Universidad Autónoma del Estado de Morelos. Cuernavaca, Morelos, México
[3]Facultad de Ciencias Biológicas, Universidad Autónoma del Estado de Morelos. Cuernavaca, Morelos, México
[4]Centro de Investigación en Biodiversidad y Conservación, Universidad Autónoma del Estado de Morelos. Cuernavaca, Morelos, México

[*] Corresponding Author's E-mail: patricia.mussali@uaem.mx.

ABSTRACT

Mining activities have more than 450 years of tradition in Mexico. This extractive activity has great relevance in the country's economy and development. However, mining activity and the processes used for the extraction of metallic minerals of economic interest generate a large amount of wastes, called mine-tailings, which contain a complex mixture of heavy metals. It has been estimated that there are many abandoned mine-tailings throughout Mexico due to the absence of legal regulations in the years before 2004, which implies a threat to surrounding ecosystems and human populations. In this chapter, a diagnosis of the current situation of the mine-tailings in the central region of Mexico is presented, 42 studies into mine tailing characterization were reviewed, published between 2005 and 2020. In these studies, a total of 27 different mine tailings were identified, possibly amounting 137 million tons of mining wastes, rich in complex mixtures of potentially toxic elements such as arsenic, cadmium, chromium, copper, iron, lead and zinc, whose bioavailability is an important concern for environmental and human health. The information presented in this chapter enables the adequate identification of these mine waste deposits, in order to evaluate the impact on human and ecological health, and to implement suitable alternatives for their environmental management.

Keywords: mining impact, environmental pollution, mine tailings

INTRODUCTION

Mining is one of the most important primary economic activities around the world. This activity includes the search, exploitation and use of the different minerals taken from nature. Mining activities directly contribute to the economic development of countries, supplying raw materials to different industries, and generating direct and indirect employments, approximately 300 million people worldwide (Cao 2007; Kossof et al. 2014). The capitalization of mineral resources contributes directly to the economic and social development of mining areas and their respective countries. However, they also generate negative impacts on the environment (Davis and Tilton 2002; Al Rawashdeh et al. 2016). Mining activities are recognized as significant causes of environmental degradation due to the release of potentially toxic elements and compounds to the environment during the different stages, from mining prospecting and exploration to the mine construction, operation, close down, or abandonment (Haddaway et al. 2019).

As a result of mining activities, large amounts of different kinds of waste are generated, including rocks and small particle size (<120 µm) materials, which are commonly known as "mine tailings." These materials are generally accumulated over land or dams at the side of the mine's installations (Jones and Boger 2012; Kossof et al. 2014; Wang et al. 2014; Söderholm et al. 2015; Baghdasaryan 2016). Mine tailings are composed of several minerals, heavy metals, mineral fuel traces and combustion ashes, fine rocks, loose sediments, and ground chemicals employed for metal extraction and recovery, as well as fluids of diverse composition (Hudson-Edwards et al. 2011).

The release of diverse heavy metals and toxic metalloids such as arsenic (As), cadmium (Cd), copper (Cu), iron (Fe), lead (Pb), zinc (Zn), among others, from the mine tailings to the environment is recognized as an ecological and human health threat (Romero et al. 2007). Due to this, an important environmental task is the adequate management of these mining wastes (Lèbre et al. 2016). In this chapter, we describe the situation relative to mine tailing deposits situated in the central region of Mexico as an attempt to evaluate the human and ecological risk as well as considering alternatives for environmental management.

THE IMPORTANCE OF MINING IN MEXICO

Mining is an important economic activity in Mexico, with more than 450 years of tradition in the country. Mining activities are related to growth and development, particularly in relation to the generation of employment, currency, and as raw material suppliers for different industrial sectors, including manufacturing, building, electric and electronics, among the most important (SGM 2018a). According to Secretary of Economy data, the mining sector in Mexico contributes four percent to the national Gross Domestic Product (SE 2021). In the

32 states of the Mexican Republic, there are metallic and nonmetallic mining deposits identified, related to the geographic locations in a volcanic region rich in different kinds of ore deposits (Volke et al. 2005; SGM 2018a).

Mexico leads mining exploration investment in Latin America and is the fourth investment destination worldwide (SE 2021). Mexico is one of the main world producers of 22 different minerals, highlighting silver (Ag, 1° place), bismuth (Bi, 3° place), molybdenum (Mo, 5° place), lead (Pb, 6° place), selenium (Se, 6° place), zinc (Zn, 6° place), cadmium (Cd, 7° place), copper (Cu, 8° place), gold (Au, 8° place), manganese (Mn, 11° place) and iron (Fe, 12° place). The extraction of economically valuable minerals takes place in mining installations and complexes know as metal processing plants. According to the Mexican Geological Service (SGM, acronym in Spanish), are 270 metal processing plants in the Mexican territory. These installations are to be found in 26 of the 32 states of Mexico (SGM 2018b).

The geographic region of Mexico with the highest number of metal processing plants is the northeastern, a region that includes the states of Chihuahua, Coahuila, Durango, Nuevo León, San Luis Potosí, Tamaulipas, and Zacatecas. It has a total of 127 metal processing plants, 91 of which are in production and 36 are inactive. The northwest region of Mexico, which includes the states of Baja California, Baja California Sur, Sinaloa, and Sonora, has a total of 80 metal processing plants, of which 52 are in production, 25 inactive, two in sporadic status and one in modernization. In the western region of Mexico, in the states of Aguascalientes, Colima, Guanajuato, Jalisco, Michoacán, Nayarit and Querétaro, there is a total of 34 metal processing plants, of which 22 are in production and 12 inactive. In the southern region of Mexico, in the states of Chiapas, Guerrero, Oaxaca and Veracruz, there are 16 mines in production. In the central region of Mexico, including Mexico City, the state of Mexico, Hidalgo, Morelos, Puebla and Tlaxcala, there are 17 metal processing plants, of which only one is inactive. Finally, the southeast region, which includes the states of Campeche, Quintana Roo, Tabasco, and Yucatan, has no history of metallic mining (Table 1).

Table 1. Metal processing plants in Mexico

		Mining complexes		
Region	States	In production	Inactive	Metal extraction
Central	Mexico city State of Mexico Hidalgo Morelos Puebla Tlaxcala	16	1	Ag, Au, Cd, Cu, Fe, Mn, Pb, Zn
Northeastern	Chihuahua Coahuila Durango Nuevo León Tamaulipas San Luis Potosí Zacatecas	91	36	Ag, Au, Ba, Cu, Fe, Mn, Pb, Zn
Northwest	Baja California Baja California Sur Sinaloa Sonora	52	25	Ag, Au, Co, Cu, Fe, Mo, Pb, Se, Zn
Southeast	Campeche Quintana Roo Tabasco Yucatan	0	0	-
Southern	Chiapas Guerrero Oaxaca Veracruz	16	0	Ag, Au, Cu, Pb, Zn
Western	Aguascalientes Colima Guanajuato Jalisco Michoacán Nayarit Querétaro	22	12	Ag, Au, Cu, Fe, Pb, Zn
	Total	196	74	

The total of active and inactive metal processing plants is shown, as well as the metals that are extracted in each region, source SGM 2018b.

Mining Wastes

As mentioned above, the mining industry plays an important economic role in Mexico. However, the adverse impacts of this industry are relevant. According to the Mexican Secretary of the Environment and Natural Resources (SEMARNAT acronym in Spanish), out of all industrial activities in Mexico, mining takes third place in terms of hazardous waste generation; 14.1% of the total generation of this kind of waste (SEMARNAT 2015). The wastes generated due to mineral extraction processes are denominated as mine tailings (Volke et al. 2005). These wastes can be defined as the residual sludge that originated during the extraction of minerals of economic value (Gómez-Bernal et al. 2010).

In their composition, mine tailings have significant amounts of sulfured compounds without economic value, such as pyrite (FeS_2), pyrrhotite ($Fe_{1-x}S$), galena (PbS), sphalerite (ZnS), chalcopyrite ($CuFeS_2$) and arsenopyrite (FeAsS), which are sources of different potentially toxic elements as heavy metals (Romero et al. 2008). The exposition of these minerals to oxygen and water generates mine acid drainage (Park et al. 2019). The low pH values, ranged from 3.5 to 5, of the mine acid drainage favoring the availability and mobility of heavy metals in mine tailings (Nordstrom 2011). Releasing and disseminating heavy metals in the surrounding environment from mine tailings generates air, soil and water pollution and negative impacts animal, plant and human populations situated in areas close to mine tailings (Plumlee and Morman 2011; Park et al. 2019).

Mine Tailing Deposits in Mexico

The extensive distribution of metal processing plants, active or inactive, in more than 80% of the states that make up Mexican territory, leads to waste being generated, known as mine tailings, and disseminated over an important area across Mexico. For example, it is estimated that in just one mining district, known as El Triunfo-San Antonio and located in the state of Baja California Sur (Northwestern region), there are around 0.8 million tons of mine tailings with high concentrations of toxic elements, including arsenic (As), barium (Ba), Cd, Fe, nickel (Ni), Pb and Zn (CENICA 2003). Another example, in the state of Guanajuato (Western region), there are 40.3 million tons of mining wastes, distributed in 31 mine tailing deposits, with considerable concentrations of Cu, Pb and Zn (Ramos-Arroyo and Siebe-Grabach, 2006).

In the state of Queretaro, there are five mine tailing deposits generated by the mining activities of the mine known as "La Negra" with high concentrations of As, Cd, Cu, Pb and Zn (Santos-Jallath et al. 2013). In the Northeastern region, in the state of Coahuila, the mining activities of just one mine known as "La Encantada" have generated six million tons of mine tailings (Rodríguez-Valles 2016). Concerning the Southern region, it has been estimated that five million tons of mine tailings, make up of heavy metals and metalloids such as Ag, As, Cd, Cu, Fe, Mn, Pb and Zn, are present in the community known as "El Fraile" in Taxco de Alarcón in the state of Guerrero (Santiago-Dionisio 2011). In the Central region of Mexico, it is estimated that in only the state of Morelos, there are around 0.78 million tons of abandoned tailings (Velasco et al. 2004), which contain heavy metals such as Cd, Cu, Fe, Mn, Pb and Zn (Solís-Miranda 2016).

INFORMATION AVAILABILITY ABOUT MINE TAILING DEPOSITS IN MEXICO

In Mexico, information related to the situation about mine tailings is not very accessible. It is scattered around in various sources (including some articles and thesis works). It is scarce and the information is not always recent. In the year 2001, 80 mine tailing deposits were estimated to be in operation (Ramírez, 2001). To date, a detailed inventory of the actual number of mine tailings (both inactive or in operation) deposits, the amount of mine wastes they accumulate, and their characteristics are not available for Mexico. Due to this, it is necessary to search for punctual information and data recompilation from a variety of different information sources. By contrast, organized and actualized information relating to the economic benefits of mining activities in Mexico is available through the official websites of the Secretary of Economy and the Mexican Geological System. This information includes the number of metal processing plants in operation, the volume and economic value of mining production, the number of employments generated, among other data, for each state of Mexico.

Mexican Normativity relation to Mining Wastes

As mentioned above, mining activity in Mexico began around 450 years ago. However, the legislation for adequate management of mining wastes was recently established. According to the Mexican General Law for the Prevention and Comprehensive Management of Waste (LGPGIR, acronym in Spanish), first published in 2003, mining wastes are considered hazardous waste due to the presence of toxic elements such as heavy metal and metalloids (Gutiérrez-Ruiz and Moreno-Turrent 2007).

The residual sludges generated as a result of the Al, Cu, Pb, and Zn production are listed as hazardous waste according to the Mexican normativity (NOM-052-SEMARNAT-2005). However, this classification is only applicable to the waste generated after normativity approval, so abandoned mine tailings constitute an environmental passive, as well as an environmental and human health threat, in several regions of Mexico (DOF 2006). Subsequently, in 2011 the normativity denominated as NOM-157-SEMARNAT-2009 was approved. This normativity establishes the elements and procedures to implement mining waste management plans for mining waste. This standard aims to minimize the generation and maximize the valuation of mining waste under criteria of environmental, technological, economic, and social efficiency (DOF 2011).

Metallic Mining Production in Central Mexico Region

Mexico is rich in metallic minerals, such as Ag, Au, Cu, Fe, Mn, Pb and Zn, and mined in much of the Mexican territory. As a result of these extractive activities, several mine tailings have been abandoned across various regions of Mexico due to the absence of legal regulations in the years before 2004. One of these regions is central Mexico, comprised of six states: Mexico City, the state of Mexico, Hidalgo, Morelos, Puebla, and Tlaxcala (Figure 1).

Currently, of the six states in Central Mexico, only three of them (state of México, Hidalgo, and Puebla) continue with metallic mining activities (SGM 2018b). Hidalgo is one of the most important mining states and in production, it is situated among the ten principal mining producers of the five metallic minerals Mn, Pb, Cu, Zn and Ag. In addition, the state of Mexico also shows important mining activity and has a significant production of Zn, Pb, Ag and Au. However, Puebla is notable only for the production of Fe, taking ninth place at the national level (SINEM 2018). The states of Mexico City, Morelos and Tlaxcala, do not show recent metallic mining activity. Tlaxcala does not have significant ore deposits on its territory and, historically, has no record of metallic mining activity. Mexico City has not reported metallic mining activities in the last 15 years. Finally, in Morelos, the metallic mining activities have been intermittent, and since 2010 no mining activities have been reported (SGM 2018b).

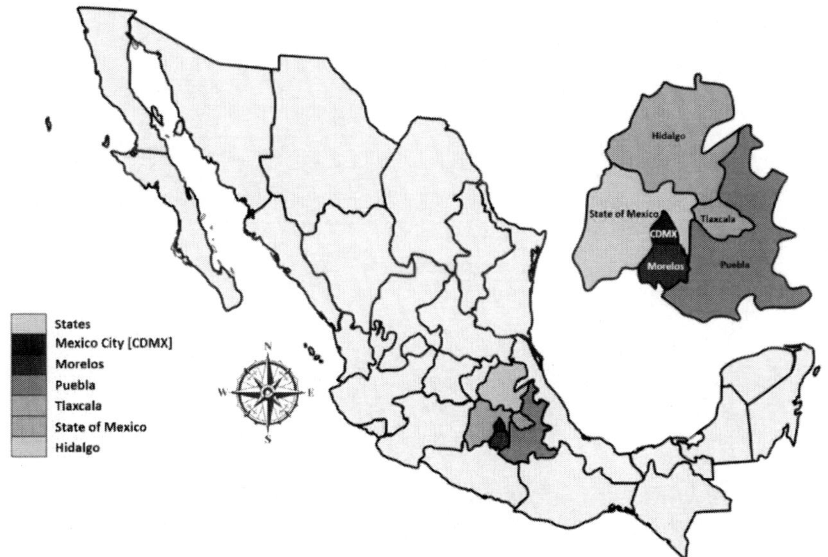

Figure 1. Central Mexico region.

Table 2. Mining projects in the central Mexico region

State	Project name	Development state	Minerals	Mining company
State of Mexico	Ancón de los Braciles	Postponed	Au	Beijing Gondwana Resources Co. Ltd. China
	Cuchara Oscar/San Ramón	In Production	Ag, Au	Impact Silver Corp. Canada
	Mirasol/San Patricio	In Production	Ag, Au	
	Zacualpan/Guadalupe	In Production	Ag, Au	
	El Oro	In exploration	Au, Ag	Candente Gold Corp. Canada
	Ixtapan del Oro	In exploration	Au, Ag	Mako Mining Corp. Canada
	Miahuatlán	Postponed	Ag	
	La Guitarra	Postponed	Au, Ag	First Majestic Silver Corp. Canada
	Mina de Agua	Postponed	Au, Ag	
	Pericones	Postponed	Ag, Au	Beijing Gondwana Resources Co. Ltd. China
	Tizapa	In Production	Zn, Pb, Ag	Dowa Mining Co. Ltd/Sumitomo Corp./Peñoles. Japan
Hidalgo	Chilcuautla y Cuautepec	In exploration	Ag, Au	Almadex Minerals Limited, Canada
	El Santuario	In exploration	Ag, Au	Palamina Corp. Canada
	La Carmen-La Joya	In exploration	Ag, Au	Plata Latina Minerals Corp. Canada
	Pachuca Real (Norte)	Postponed	Ag, Au	Solitario Zinc Corp. /Hochschild Mining Plc. USA/Peru
	Pachuca SE	In exploration	Ag, Au	Prospero Silver Corp. Canada
	Petate	In exploration	Au, Ag	
	Pisaflores	Postponed	Au, Ag, Cu, Pb, Zn	China Minerals Resources Group. China
Morelos	Corazonada y San José	Postponed	Ag, Pb, Zn	Minaurum Gold Inc. Canada
	Esperanza (Cerro Jumíl)	In development	Au, Ag, Cu	Alamos Gold Inc. Canada
	Mercury Mines	In exploration	Ag, As, Hg	
	Temixco	Postponed	Au, Ag	Gunpower Capital Corp. Canada
Puebla	Caldera	In exploration	Au, Ag	Almadex Minerals. Canada
	Cuyoaca	Postponed	Au, Ag, Cu	
	La Preciosa	In exploration	Au, Ag	

State	Project name	Development state	Minerals	Mining company
Puebla	Cerro Dolores	In exploration	Ag, Pb, Zn	Starcore International Mines Ltd/Goldcorp Inc. Canada
	Ixtaca (Tuligtic)	In development	Au, Ag	ARGC Mining Company/JC Mining. China/Peru
	La Lupe	Postponed	Au	Chesapeake Gold Corp. Canada
	Nueva España	In exploration	Au, Ag	Minaurum Gold Inc. Canada
	Pórfido	In exploration	Au	Almaden Minerals Ltd. Canada
	Rosa	In exploration	Au, Cu	
	El Chato	In exploration	Au, Ag	
	Zapotec	In exploration	Au, Ag	Oro East Mining Inc. USA

In addition to mines in production, several mining projects have been proposed for the central Mexico region. In the year 2018, 34 mining projects for metallic mineral extraction were proposed for Estado de México, Hidalgo, Morelos, and Puebla (Table 2). These mining projects are in different stages of development and according to the information for 2018, 17 mining projects were in the exploratory stage, 11 were postponed due to socio-environmental problems and/or issues related to economic capital, four were in the production stage and two more in the development stage (SINEM 2019). It is important to mention that 82.4% of the mining projects reported in 2018 for the central Mexico region belong to Canadian mining companies. The remaining 7.6% belong to mining companies from five different countries: China, Japan, Mexico, Peru, and the United States. Currently, most of the mining projects in Mexico are in the hands of foreign companies.

Metal Mines in the Central Mexico Region

The great mining activity in the central region of Mexico has led to the exploitation of several metallic mineral mines throughout the mining history of the country. The Mexican Geological System has been in charge of registering all mines located in Mexican territory since 1995, and this information is available through the GeoInfoMex consultation system.

Table 3. Number of metal mines situated in central Mexico region

	State			
Mine status	**State of Mexico**	**Hidalgo**	**Morelos**	**Puebla**
Abandoned	97	282	17	113
In situ mining	47	80	20	180
Prospect	46	65	5	38
In production	8	12	0	9
Unknown	1	1	0	1
Total	199	440	42	341

Mexico City and Tlaxcala do not report metal mining activities, self elaboration based on data from the SGM (2020).

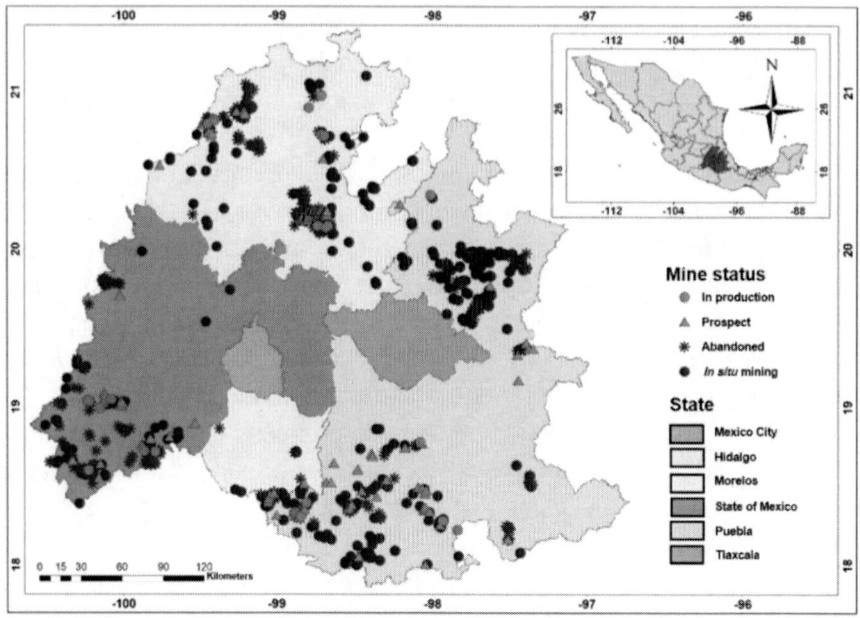

Figure 2. Distribution of metallic mines in Central Mexico region.

According to the information from the GeoInfoMex consultation system, this region has a total of 1,022 metal mines (Figure 2), most of them established in the mining districts of Hidalgo (Table 3), where there are a total of 440; the state of Puebla with a total of 341 mines, while the state of Mexico has a total of 199 mines. Finally, the state of Morelos has a total of 42 mines in its territory, most of which are abandoned. Thus, there are only 29 mines in production, but the small-scale *in situ* mining activities have taken on great relevance in the region (SGM 2020).

Classification of Metal Mines in the Central Mexico Region

According to the SGM (2020), the mines are classified according to the status they present as abandoned, in production, prospect, *in situ* mining and unknown status. In the central region of Mexico, there is a higher proportion of abandoned and *in situ* mines (81.8%). The abandoned mines showed metallic production, however, they do not report production currently, these mines represent the 49.8% of the total mines registered for the region. On the other side, *in situ* mines report small-scale operations; these mines represent 32% of the total mines registered for the region. In these mines, metals are extracted through chemical lixiviation; this kind of operation does not generate the impacts commonly associated with high-scale mining activities, such as excessive noise, atmospheric contamination by dust, or mine tailings generation (Ortiz-Sánchez et al. 2010). Finally, the in prospect mines are sites with potential for implementation of mining activities, the number of these sites represent 15.7% of the total mines registered for the region, while only 2.8% of the mines of the region are currently in production (SGM 2020).

In the present chapter, the abandoned and in operation mines are of particular interest since they have generated or are still generating waste (mine tailings) during the processes for metal production. In abandoned mines, the waste generated when they were operating is also abandoned close to the mines. This waste is generally accumulated in the open, without treatment allowing the mobilization of toxic elements due to weather phenomena such as climate and hydrology (Plumlee and Morman 2011). The abandoned and in production mines represent 52.6% of the total number of mines situated in the central region of Mexico. Therefore, there must be a significant number of abandoned tailings in the areas close to these mines.

STATUS OF METAL MINES IN CENTRAL MEXICO REGION

The mines situated in the central region of Mexico have registered the extraction of different metallic minerals; Ag is the principal element found in most of the registered mines (abandoned or in operation). In the state of Mexico, Ag production was reported in 95.2% of the mines, in the state of Morelos in 94.1%, 85.7% in Hidalgo and 58% in Puebla. However, the extraction of other economically valuable metals has been reported, and the extraction of Au was reported in 77.5% of the mines of Hidalgo, 62% in the state of Mexico, and 57.5% in Puebla. The extraction of Zn was reported in 64.7% of the mines of Morelos and in 38.6% of the mines of Puebla. The extraction of Pb was reported for 53% of the mines of Morelos, 39.1% of the mines of the state of México and 17% of the mines of Hidalgo. Generally speaking, the main metals extracted in the 538 mines (abandoned or in operation) situated in the region were Ag (82.5%), Au (69.1%), Pb (26%) and Zn (24.7%). In this area, most of the mines registered are abandoned (94.6%). Currently, in the state of Morelos, all reported mines are abandoned, while in the state of Hidalgo, 96% of the mines have been abandoned, 92.6% in Puebla and 92.4% in the state of Mexico.

MINE TAILING DEPOSITS IN THE CENTRAL MEXICO REGION

In the present chapter, we have identified 41 studies (articles and thesis) covering the period from 1995-2020, in which the characterization of mine tailing deposits is described (Table 4). Twenty-four of these studies were carried out in mine tailing deposits situated in Hidalgo, ten corresponding to the characterization of mine tailing deposits in the state of Mexico, and seven for the state of Morelos. In the reports for Hidalgo, 19 mine tailing deposits were evaluated, these waste deposits were situated in three municipalities: Zimapán, Omitlán de Juárez and Pachuca. However, according to the geographical coordinates and description, several mine tailing deposits were referred to with distinct names in the studies. In the studies for the state of Mexico, five mine tailing deposits were evaluated, located in two municipalities: Zacazonapan and El Oro de Hidalgo. In the state of Morelos, only three mine tailing deposits were evaluated. It is important to mention that in the state of Puebla, despite being the second state with the number of metallic mines reported for the region, there were no identified studies relative to mine tailing deposits characterization for the information search period.

Mine Tailing Amounts in the Central Mexico Region

It has been estimated that the municipality of Pachuca, in the state of Hidalgo, is home to 100 million tons of mining waste, occupying 286 hectares (Patiño et al. 2016; Volpi-León 2017), of which a little more than 14.3 million tons are from the mine tailing deposit called Dos Carlos, arranged in an area of 23 hectares (Hernández-Acosta et al. 2009; Juárez-Tapia et al. 2018). Together, the municipalities of Pachuca and Omitlán de Juárez present a total of 108.1 million tons of tailings in their territory, occupying a total area of 1,200 hectares (Flores et al. 2015). According to the collected information, 19 mine tailing deposits were identified in the state of Hidalgo, which amounts to 108.1 million tons of mine tailings distributed in six deposits in the municipalities of Pachuca and Omitlán de Juárez. The remaining 13 mine tailing deposits were located in the municipality of Zimapán, however, there is no estimate of the amount of mining waste deposited in this municipality.

Table 4. Characteristics of mine tailing deposits in the Central Mexico region

<table>
<tr><th rowspan="2">Municipality</th><th rowspan="2">Mine tailing name</th><th colspan="2">Location</th><th rowspan="2">Present elements</th><th colspan="3">Characteristics</th><th rowspan="2">Reference</th></tr>
<tr><th>Latitude (N)</th><th>Longitude (W)</th><th>pH</th><th>EC (dS/m)</th><th>OM (%)</th></tr>
<tr><td colspan="9">State of Mexico</td></tr>
<tr><td rowspan="5">El Oro de Hidalgo</td><td>El Oro 1</td><td>19.812</td><td>-100.136</td><td rowspan="2">Ag, As, Au, Cd, Co, Cr, Cu, Ni, Pb, Zn</td><td>4.5-8.5</td><td>1.9</td><td>-</td><td rowspan="2">Maldonado-Villanueva 2008</td></tr>
<tr><td>El Oro 2</td><td>19.811</td><td>-100.123</td><td>7.4-8.8</td><td>1.6</td><td>-</td></tr>
<tr><td>Tiro México</td><td>19.812</td><td>-100.136</td><td>Ag, As, Co, Cr, Cu, Pb, Ni, V, Zn</td><td>4.5-8.5</td><td>0.2-0.9</td><td>-</td><td>Corona-Chávez et al. 2010</td></tr>
<tr><td>Conalep</td><td>19.811</td><td>-100.123</td><td rowspan="2">Ag, As, Au, Be, Cd, Co, Cr, Cu, Ni, Pb, Rb, Sb, V, Zn</td><td>8.7</td><td>0.9</td><td>-</td><td rowspan="2">Corona-Chávez et al. 2017</td></tr>
<tr><td>Tiro México</td><td>19.812</td><td>-100.136</td><td>7.9</td><td>0.6</td><td>-</td></tr>
<tr><td rowspan="10">Zacazonapan</td><td rowspan="2">Tizapa</td><td>-</td><td>-</td><td>Ag, As, Cd, Cu, Fe, Ni, Pb, Zn</td><td>6.0</td><td>-</td><td>-</td><td>Santos-Martínez 2006</td></tr>
<tr><td>-</td><td>-</td><td>As, Cu, Fe, Ni, Pb, Zn</td><td>5.8</td><td>2.3</td><td>-</td><td>Flores-Alvares 2008</td></tr>
<tr><td rowspan="4">Presa de jale</td><td>-</td><td>-</td><td>Ag, As, Ba, Cd, Cu, Fe, Li, Mn, Mo, Ni, Pb, Zn</td><td>6.8- 7.4</td><td>4.3-6.3</td><td>-</td><td>González-Sandoval et al. 2008</td></tr>
<tr><td>-</td><td>-</td><td>As, Cu, Fe, Ni, Pb, Zn</td><td>5.8</td><td>2.3</td><td>-</td><td>Zamora-Duarte 2008</td></tr>
<tr><td>19.035</td><td>-100.235</td><td>As, Ba, Cu, Fe, Pb, Zn</td><td>-</td><td>-</td><td>-</td><td>Lizárraga-Mendiola et al. 2008</td></tr>
<tr><td>19.035</td><td>-100.235</td><td>Ag, Al, As, Au, Ca, Cu, Fe, K, Mg, Mn, Pb, Si, Zn</td><td>-</td><td>-</td><td>-</td><td>Lizárraga-Mendiola et al. 2009</td></tr>
<tr><td>Activa</td><td>19.035</td><td>-100.245</td><td>Ag, As, Ba, Cd, Cu, Pb, Se, Zn</td><td>4.0-5.3</td><td>1.7-3.3</td><td>-</td><td rowspan="2">Martínez-Soriano 2011</td></tr>
<tr><td>Inactiva</td><td>19.036</td><td>-100.241</td><td>As, Ag, Ba, Cu, Pb, Zn</td><td>2.9-6.4</td><td>2.0-3.9</td><td>-</td></tr>
<tr><td>Tizapa-jale I</td><td>19.035</td><td>-100.245</td><td>Si, Zn</td><td>7.7</td><td>2.6</td><td>-</td><td rowspan="2">Flores-Mendoza 2013</td></tr>
<tr><td>Tizapa-jale II</td><td>19.036</td><td>-100.241</td><td>Ag, Al, As, Ba, Cd, Cu, Fe, Pb, Se, Si, Zn</td><td>6.4</td><td>0.3</td><td>-</td></tr>
<tr><td colspan="9">Hidalgo</td></tr>
</table>

Municipality	Mine tailing name	Location		Present elements	Characteristics			Reference
		Latitude (N)	Longitude (W)		pH	EC (dS/m)	OM (%)	
Omitlán de Juárez	Presa Velasco	20.179	-98.647	Ag, As, Au, Ba, Be, Cd, Co, Cr, Cu, Mo, Ni, Pb, Sb, Zn	-	-	-	Flores et al. 2015
Pachuca	Dos Carlos	20.105	-98.713	Cd, Cu, Mn, Ni, Pb, Zn	7.0-7.7	-	0.07-0.14	Hernández-Acosta et al. 2009
		-	-	Ag, Al, Au, Bi, Ca, Cu, Fe, K, Mg, Mn, Na, Ni, P, Si, Sn, Ti,	-	-	-	Patiño et al. 2016
		-	-	Al_2O_3, CaO, Fe_2O_3, K_2O, MnO, Na_2O, P_2O_5, SiO_2, SO_3, TiO_2	-	-	-	Juárez-Tapia et al. 2018
Pachuca-Real del Monte	Dos Carlos	20.105	-98.713	As, Cd, Cu, Fe, Mn, Ni, Pb, Zn	7.3-7.9	0.2-0.3	-	Lizárraga-Mendiola et al. 2014
	Dos Carlos	20.105	-98.713	Ag, As, Au, Ba, Be, Cd, Co, Cr, Cu, Mo, Ni, Pb, Sb, Zn	-	-	-	Flores et al. 2015
	Santa Julia Sur o Sur II	20.076	-98.761		-	-	-	
	Santa Julia Norte o Sur I	20.091	-98.757		-	-	-	
	Dos Carlos	20.105	-98.756	Ag, Al_2O_3, Au, CaO, Fe_2O_3, K_2O, MgO, MnO, Na_2O, P_2O_5, SiO_2, TiO_2	-	-	-	Volpi-León 2017
	Presa Sur	20.081	-98.765		-	-	-	
Zimapán	San Miguel viejo 1	20.727	-99.397	As, Cd, Cr, Cu, Fe, Ni, Pb, Zn	2.1-2.9	2.5-5.4	-	Pérez-Martínez 2005
	San Miguel viejo 2				6.2-8.2	2.3	-	
	San Miguel	20.736	-99.399	As, Cd, Cr, Hg, Pb	7.3- 7.5	-	-	Hernández-Zamora 2009
	Santa María	20.727	-99.397		6.5- 7.5	-	-	
	El Monte	20.825	-99.372	Ag, As, Bi, Cd, Cr, Cu, Ni, Pb, Sb, Se, Zn	-	-	-	Moreno-Tovar et al. 2009
	San Miguel	-	-		-	-	-	
	Santa María	20.736	-99.399		-	-	-	

Table 4. (Continued)

Municipality	Mine tailing name	Location		Present elements	Characteristics			Reference
		Latitude (N)	Longitude (W)		pH	EC (dS/m)	OM (%)	
Zimapán	Pal	20.726	-99.383	As, Cu, Fe, Ni, Pb, Zn	2.1	-	13.6	Hernández-Ruiz-Gaytán and Padilla-Carrillo 2010
	Preisser	20.727	-99.384		2.1-2.2	-	14.6	
	San Francisco [Gray Non Ox]	20.727	-99.397		7.6	-	2.6	
	San Francisco [Red Ox]				2.05-2.74	-	3.6-14.49	
	No 9	20.731	-99.396	As, Cd, Cu, Fe, Mn, Pb, Rb, Sr, Zn, Zr	6.9-7.37	0.43	-	Guzmán-Rangel 2012
	El Monte	20.825	-99.372	As, Bi, Cd, Cr, Cu, Fe, Ni, Pb, Sb, Se, Zn	-	-	-	Moreno-Tovar et al. 2012
	San Miguel	-	-		-	-	-	
	Santa María	20.736	-99.399		-	-	-	
	Lomo de Toro	20.727	-99.382	Cd, Cu, Mn, Ni, Pb, Zn	7.1	2.2	-	Duarte-Zaragoza et al. 2014
	Los Gómez	20.727	-99.397		1.7	20	-	
	Pal Nueva	20.726	-99.383		6.3	1.9	-	
	Preisser	20.727	-99.384		3.0	12.4	-	
	San Antonio	20.727	-99.381		7.8	1.4	-	
	San Miguel Viejo	20.727	-99.391		7.5	1.8	-	
	Santa María	20.736	-99.399		7.8	1.3	-	
	Zimapán Co	20.726	-99.385		1.7	11.1	-	
	Gray mining tailings	-	-	As, Cd, Fe, Pb, Zn	6.0-6.1	-	-	Zúñiga-Vázquez 2014
	Red mining tailings	-	-		2.4	-	-	
	San Miguel Viejo	20.727	-99.397	As, Cd, Cu, Fe, Mn, Pb, V, Zn	Dry: 6.9-8.2 Rains: 5.5-6.3	-	-	Armienta et al. 2016

Municipality	Mine tailing name	Location		Present elements	Characteristics			Reference
		Latitude (N)	Longitude (W)		pH	EC (dS/m)	OM (%)	
Zimapán	San Miguel Nuevo	20.736	-99.399		Dry: 7.2-8.4 Rains: 5.4-6.4	-	-	
	Red mining tailings	-	-		Dry: 2.0-7.6 Rains: 2.2-6.1	-	-	
	San Francisco [Gray Non Ox]	20.825	-99.372	Cd, Ni, Pb	7.8	1.6-4.1	8.2	Sánchez-López et al. 2015
	San Francisco [Red Ox]				1.7	5-80	8.2	
	Santa María	20.736	-99.399		7.6	4	1.1-8.2	
	San Miguel Viejo	20.727	-99.390	As, Cd, Fe, Pb, Zn	7.5	5.7-8.5	5.1-5.2	Armienta et al. 2019
	No 5	-	-	Cu, Pb, Zn	2.2	0.7	-	Cruz-Ruiz 2017
	No 9	-	-		2.2	0.7	-	
	Santa María	20.736	-99.399	Cu, Cd, Pb, Zn	6.0	2.5	0.2	González-Chávez et al. 2017
	No 9	-	-	As, Ca, Cu, Fe, K, Mo, Ni, Pb, Rb, S, Sb, Sn, Sr, Ti, Zn, Zr	7.8-8.4	-	-	García-Luna 2018
	No 9	-	-	As, Pb	7.54	-	-	Gutiérrez-Bazán 2018
	CMZ/Pal	20.727	-99.397	As, Cu, Cd, Mn, Pb, Sb, Se, Zn	2.5	4.68	-	Molina-Garduño 2018
	El Espíritu	20.727	-99.399		7.9	1.202	-	
	El Espíritu reciente	20.728	-99.398		8.4	1.153	-	
	San Miguel Nuevo	20.736	-99.399		4.91	2.26	-	
	San Francisco	20.831	-99.374	As	-	-	-	Franco-Farfán 2019

Table 4. (Continued)

Municipality	Mine tailing name	Location		Present elements	Characteristics			Reference
		Latitude (N)	Longitude (W)		pH	EC (dS/m)	OM (%)	
Zimapán	A	-	-	As, Cd, Fe, Pb, Zn	-	-	-	Zúñiga-Vázquez 2019
	B	-	-		-	-	-	
	C	20.736	-99.399		-	-	-	
Morelos								
Tlaquiltenango	A	18.438	-99.031	Cd, Cr, Ni, Pb	7.74	-	1.18	Hernández-Zamora 2009
	B	18.437	-99.031		7.12	-	0.59	
	Mining tailing 1	18.434	-99.022	As, Cd, Mn, Pb, V	8.2	-	-	Tovar-Sánchez et al. 2012
	Mining tailing 2	18.437	-99.031		8.2	-	-	
	Mining tailing 3	18.443	-99.024		8.2	-	-	
	Mining tailing 1	18.434	-99.022	As, Cd, Mn, Pb	8.2	-	-	Mussali-Galante et al. 2013
	Mining tailing 2	18.437	-99.031		8.2	-	-	
	Mining tailing 3	18.443	-99.024		8.2	-	-	
	Mining tailing	18.434	-99.022	As, Cu, Pb, Zn	-	-	-	González-Brito 2015
	Mining tailing 1	18.434	-99.022	Cu, Cd, Fe, Mn, Pb, Zn	8.1-8.4	0.4	0.8-0.9	Solís-Miranda 2016
	Mining tailing 2	18.437	-99.031		7.5-8.5	0.1-0.7	0.5-0.6	
	Mining tailing	-	-	Ag, Al, As, Au, Ca, Cu, Fe, Mn, Si, Ti, Zn	-	-	-	Martínez-Soto et al. 2020
	Mining tailing 1	18.434	-99.022	Cu, Fe, Pb, Zn	-	-	-	Santoyo-Martínez et al. 2020
	Mining tailing 2	18.437	-99.031		-	-	-	
	Mining tailing 3	18.443	-99.024		-	-	-	

In the state of Mexico, the amount of mine tailings has been estimated at 28.5 million tons, of which 23 million tons are located in the municipality of El Oro de Hidalgo, while 5.5 million tons are located in the municipality of Zacazonapan. This estimate corresponds to five mine tailing deposits situated in the state of Mexico. In state of Morelos, a total amount of 0.78 million tons has been estimated, distributed in three mine tailings deposits in the community of Huautla in the municipality of Tlaquiltenango (Hernández-Zamora 2009; Tovar-Sánchez et al. 2012; Mussali-Galante et al. 2013; Solís-Miranda 2016; Santoyo-Martínez et al. 2020).

According to the studies identified, in the central region of Mexico, there are 27 mine tailing deposits (Figure 3). However, a total of 137.4 million tons of mining waste has been estimated according to the information corresponding to 14 mine tailing deposits. As a result, there is a significant amount of mining waste not contemplated in these estimates.

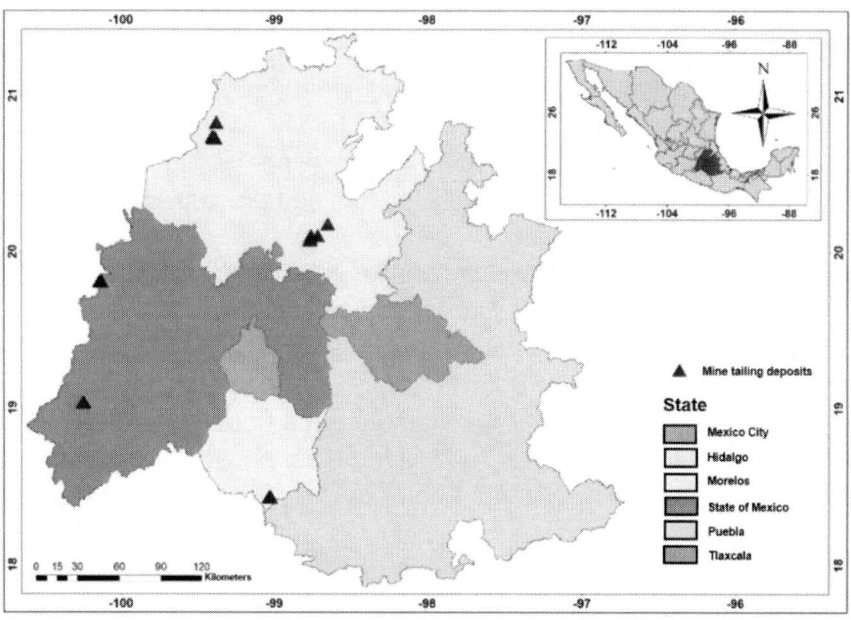

Figure 3. Mine tailing deposits in Central Mexico region.

CHARACTERIZATION OF MINE TAILING IN THE CENTRAL MEXICO REGION

According to the information collected from the studies, the chemical composition of mine tailings includes a complex mixture of different heavy metals and metalloids, including Ag, Al, As, Au, Cd, Cr, Cu, Fe, Mn, Ni, Pb and Zn. While, As, Cu, Fe, Pb and Zn were the most reported in the mine tailings of the state of Mexico, Hidalgo, and Morelos. In addition to the state of Puebla, these states report the presence of hydrothermal ore deposits in their territories, composed of mixtures of the different metals Ag, Au, Cu, Fe, Mn, Pb and Zn (SGM 2020). Due to this and levels of the mining activities, it is highly probable that there exist non-reported or studied mine tailing deposits in the state of Puebla.

The oxidation processes of the metallic sulfides present in the mine tailings generate the release of acid mine drainage. Due to its low pH, the dissolution and mobilization of heavy metals from mine tailings to the surrounding environment are favored (Park et al. 2019). The acid mine drainage can also react in nonmetallic minerals as carbonate minerals, aluminum hydroxides, iron oxyhydroxides and aluminosilicates, also present in the mine tailings, generating neutralization, absorption and ionic interchange processes that limit the mobilization of heavy metals.

In the mine tailing characterization, it is important to determine parameters such as pH and neutralization potential to establish the feasibility of mobilization and dangerousness of the heavy metals present in mine tailings (Dold 2010). However, in the mine tailings characterization studies, these parameters are not always determined, reporting only the chemical composition of the mine tailings (Lizárraga-Mendiola et al. 2008; Lizárraga-Mendiola et al. 2009; Moreno-Tovar et al. 2009; Moreno-Tovar et al. 2012; Flores et al. 2015; González-Brito 2015; Patiño et al. 2016; Volpi-León 2017; Juárez-Tapia et al. 2018; Franco-Farfán 2019; Martínez-Soto et al. 2020).

The mine tailing deposits reported for the Estado de México have registered pH values in a range between 2.9-8.8, according to Mexican normativity (NOM-021-SEMARNAT-2000). Thus, these mine tailings have pH values from highly acidic to highly alkaline. Highly acidic mine tailings are situated in the municipality of the Oro de Hidalgo, while the highly alkaline mine tailings are found in the municipality of Zacazonapan (Maldonado-Villanueva 2008; Corona-Chávez et al. 2010; Martínez-Soriano 2011). With respect to the pH values reported for the mine tailings of Hidalgo, these show values between 1.7-8.4, and the more acidic mine tailings are located in the municipality of Zimapán. These mine tailings show the lowest pH values of all presents in the central region of Mexico (Pérez-Martínez 2005; Hernández-Ruiz-Gaytán and Padilla-Carrillo 2010; Duarte-Zaragoza et al. 2014; Zúñiga-Vázquez 2014; Sánchez-López et al. 2015; Armienta et al. 2016; Cruz-Ruiz 2017; Molina-Garduño 2018). Finally, the mine tailings in the state of Morelos showed moderate alkaline pH values from 7.1-8.4 (Hernández-Zamora 2009; Tovar-Sánchez et al. 2012; Mussali-Galante et al. 2013; Solís-Miranda 2016).

Conclusion

Metallic mining activity in central Mexico has taken place in Hidalgo, the state of México, Puebla, and Morelos. In these states, 1,022 metal mines are distributed, of which a little more than half could have or are generating tailings, 49.8% are abandoned mines and 2.84% are mines currently in operation. However, the state of Morelos is not producing metallic minerals at the moment, reporting only intermittent mining activity; the last activity report took place in 2010. Likewise, Tlaxcala and Mexico City have not reported deposits of metallic minerals in their territory; therefore, they do not present metallic mining activity.

A total of 42 tailings characterization studies were found, where information from 27 different tailings was obtained. Of these tailings, 19 are found in the municipalities of Zimapán, Omitlán de Juárez and Pachuca in the state of Hidalgo, five in the municipalities of Zacazonapan and Oro de Hidalgo in the state of Mexico, and three more in the municipality of Tlaquiltenango in the state of Morelos. For the state of Puebla, no studies were found on tailings, although it has a record of 122 mines (in production and abandoned) and is relevant in the production of Fe, Pb, Au, and Ag.

A total of 137.38 million tons of mining waste has been estimated in the central region of Mexico. However, according to the available information, this estimate belongs to only 51.8% of the reported mine tailings. The total amount of this mining wastes in the region could not be estimated since no information was found for the municipality of Zimapán, Hidalgo, which represents 48.1% of the mine tailing deposits reported. Specifically, for the state of Hidalgo, 100 million tons of tailings have been estimated in Pachuca and 8.1 million tons more in Omitlán de Juárez, belonging to six tailings located in these municipalities. For the state of Mexico, 5.5 million tons of tailings have been estimated in Zacazonapan and 23 million tons in El Oro de Hidalgo that belong to five tailings located in said municipalities. For the state of Morelos, a total of 780 thousand tons has been estimated, belonging to three tailings located in the municipality of Tlaquiltenango.

It is important to mention that in addition to the 27 different tailings identified, a report of another 11 tailings was found. However, these could not be counted due to the lack of information indicating their precise location. The report of the mine tailings coordinates is an essential part of their identification process because the same tailings are named differently. Without the coordinates, it is difficult to identify if it is already reported in other studies. On the other hand, studies show that all mine tailings present concentrations of a complex mixture of heavy metals in their composition, with Pb, Cu, Zn, As and Fe reported in a higher proportion. Hidalgo, Morelos, the state of Mexico and Puebla present hydrothermal type deposits with mineralization of polymetallic veins of Au, Ag, Pb, Zn, Cu, Fe and Mn. Therefore, in the state of Puebla, there could also be mine tailings with high content of Pb, Cu, Zn, As and Fe, mainly.

The pH reported in the different studies is highly variable since it goes from strongly acidic to strongly alkaline. However, the mine tailings that reported the more acidic pH were the three tailings located in the municipality of Zimapán in Hidalgo, with a pH value of 1.7. Six more mine tailings also reported a strongly acidic pH (2.13 to 4.91), they are located in Zimapán, Hidalgo, and in the localities El Oro de Hidalgo and Zacazonapan in the state of Mexico. The pH values determination is important because, in a more acidic pH, there is a greater probability of heavy metals mobilization. Despite the importance of pH in determining the degree of danger of tailings, not all studies report it.

Adding more parameters in the tailings characterization studies, and their geographic coordinates, would allow obtaining a more precise tailings inventory in terms of number and risk. In addition, it would allow the faster and easier location of the areas that are at risk from mining waste and propose alternatives to mitigate the environmental and social impacts that tailings present.

References

Al Rawashdeh, R., Campbell, G. and Titi, A. 2016. "The socio-economic impacts of mining on local communities: The case of Jordan." *The Extractive Industries and Society* 3 (2):494-507.

Armienta, M., Beltrán, M., Martínez, S. and Labastida, I. 2020. "Heavy metal assimilation in maize (*Zea mays* L.) plants growing near mine tailings." *Environmental Geochemical Health* 42 (8):2361-2375.

Armienta, M.A., Mugica, V., Reséndiz, I. and Gutierrez-Arzaluz, M. 2016. "Arsenic and metals mobility in soils impacted by tailings at Zimapán, México." *Journal of Soils and Sedimets* 16 (4):1267-1278.

Baghdasaryan, T. 2016. "*Assessment of the environmental impact of tailings in the Republic of Armenia*." Master Thesis. Universidade da Coruña.

Cao, X. 2007. "Regulating mine land reclamation in developing countries: The case of China." *Land Use Policy* 24 (2):472-483.

Centro Nacional de Investigación y Capacitación Ambiental. 2003. *Informe final: Remediación de sitios contaminados por metales Provenientes de Jales Mineros en los Distritos de El Triunfo-San Antonio y Santa Rosalía, Baja California Sur. Dirección de Investigación en Residuos y Proyectos Regionales.* Accessed July 7, 2021. http://defiendelasierra.org/wp-content/uploads/Estudio-de-Remediaci%C3%B3n-El-Triunfo-y-San-Antonio-2003.pdf. [*Final report: Remediation of sites contaminated by metals from Mine Tailings in the Districts of El Triunfo-San Antonio and Santa Rosalía, Baja California Sur. Directorate of Research on Waste and Regional Projects*]

Corona-Chávez, P., Maldonado, R., Ramos-Arroyo, Y. R., Robles-Camacho, J., Lozano-SantaCruz, R. and Martínez-Medina, M. 2017. “Geoquímica y mineralogía de los jales del distrito minero Tlalpujahua-El Oro, México, y sus implicaciones de impacto ambiental.” *Revista Mexicana de Ciencias Geológicas* 34 (3):250-273. ["Geochemistry and mineralogy of the tailings of the Tlalpujahua-El Oro mining district, Mexico, and its implications for environmental impact." *Mexican Journal of Geological Sciences*]

Corona-Chávez, P., Uribe-Salas, J. A., Razo-Pérez, N., Martínez-Medina, M., Maldonado-Villanueva, R., Ramos-Arroyo, Y. R. and Robles-Camacho, J. 2010. “The impact of mining in the regional ecosystem: the Mining District of El Oro and Tlalpujahua, Mexico.” *De re metallica (Madrid): Revista de la Sociedad Española para la Defensa del Patrimonio Geológico y Minero* 15:21-34. [*De re metallica (Madrid): Magazine of the Spanish Society for the Defense of the Geological and Mining Heritage*]

Cruz-Ruiz, E. 2017. *Efecto del encalado, abonado y fertilización de jales mineros ácidos sobre la biodisponibilidad química y bioacumulación de Cu, Cd, Pb y Zn en Festuca arundinacea.* Master Thesis. Autonomous National University of Mexico. [*Effect of liming, fertilizing and fertilization of acid mining tailings on the chemical bioavailability and bioaccumulation of Cu, Cd, Pb and Zn in Festuca arundinacea.*]

Davis, G. A. and Tilton, J. E. (2002). “Should developing countries renounce mining. A Perspective on the Debate.” *International Council on Mining and Metals (ICMM). London.* Accessed July 7, 2021. https://citeseerx.ist.psu.edu/viewdoc/download?doi=10.1.1.468.8759&rep=rep1&type=pdf.

Diario Oficial de la Federación (2006). *Norma oficial Mexicana NOM-052-SEMARNAT-2005. Que establece las características, el procedimiento de identificación, clasificación y listado de los residuos peligrosos.* Accessed July 7, 2021. http://www.economia-noms.gob.mx/normas/noms/2006/052semarnat.pdf. [*Official Mexican Standard NOM-052-SEMARNAT-2005. That establishes the characteristics, the procedure for identification, classification and listing of hazardous waste.*]

Diario Oficial de la Federación (2011). *Norma oficial Mexicana NOM-157-SEMARNAT-2009, Que establece los elementos y procedimientos para instrumentar planes de manejo de residuos mineros.* Accessed July 7, 2021. https://www.profepa.gob.mx/innovaportal/file/6665/1/nom-157-semarnat-2009.pdf. [*Official Mexican Standard NOM-157-SEMARNAT-2009, which establishes the elements and procedures to implement mining waste management plans.*]

Dold, B. 2010. "Basic concepts in environmental geochemistry of sulfidic Mine-Waste management." In *Waste Management*, edited by Kumar, E. S. 173–198. *IntechOpen,* doi:10.5772/8458.

Duarte-Zaragoza, V. M., Gutiérrez-Castorena, E., Gutiérrez-Castorena, M. C., Carrillo-González, R., Ortiz-Solorio, C. A. and Trinidad-Santos, A. 2014. "Heavy metals contamination in soils around tailing heaps with various degrees of weathering in Zimapán, Mexico." *International Journal of Environmental Studies* 72 (1):24-40.

Flores, J., Hernández, J., Pérez, M., Patiño, F., Abacú-Ostos, J. and Trápala, N. Y. 2015. "Developing alternative industrial materials from mining waste." In *Engineering Solutions for Sustainability: Materials and Resources II,* edited by Fergus, J. W., Mishra, B., Anderson, D., Sarver, E. A. and Neelameggham, N. R. 121-272. John Wiley & Sons.

Flores-Álvarez, J. F. 2008. *Pruebas de intemperismo acelerado en clumnas (celdas húmedas) de jales de minera Tizapa con microorganismos mesófilos tipo Acidithiobacillus ferrooxidans.* Bachelors Thesis. Autonomous National University of Mexico. [*Accelerated weathering tests in columns (wet cells) of Tizapa mining tailings with mesophilic microorganisms type Acidithiobacillus ferrooxidans*]

Flores-Mendoza, J. A. 2013. *Pruebas estáticas y cinéticas para valuar peligrosidad de jales.* Bachelors Thesis. Autonomous National University of Mexico. [*Static and kinetic tests to assess the danger of tailings*]

Franco-Farfán, E. F. 2019. *Estudio de especiación de arsénico por EAA en jales mineros del estado de hidalgo, México y su biodisponibilidad.* Bachelors Thesis. Autonomous National

University of Mexico. [*Study of arsenic speciation by EAA in mining tailings of the state of Hidalgo, Mexico and its bioavailability.*]

García-Luna, N. 2018. *Evaluación de la ecotoxicidad de lixiviado de jales utilizando embriones de Danio rerio como organismo de prueba.* Master Thesis. Autonomous National University of Mexico. [*Evaluation of tailings leachate ecotoxicity using Danio rerio embryos as test organism.*]

Gómez-Bernal, J. M., Santana-Carillo, J., Romero-Martin, F., Armienta-Hérnández, M. A., Morton-Bermea, O. and Ruiz-Huerta, E. A. 2010. "Plantas de sitios contaminados con desechos mineros en Taxco, Guerrero, Mexico." *Boletín de la Sociedad Botánica de México* 87:131-133. ["Plants of sites contaminated with mining waste in Taxco, Guerrero, Mexico." *Bulletin of the Botanical Society of Mexico*]

González-Brito, W. A. 2015. *Efecto de un gradiente de contaminación por jales mineros sobre las comunidades de artrópodos asociados a la vegetación y con énfasis en el orden Araneae en Huautla, Morelos.* Bachelors Thesis. Autonomous University of Morelos State. [*Effect of a pollution gradient by mining tailings on the arthropod communities associated with vegetation and with emphasis on the order Araneae in Huautla, Morelos.*]

González-Chávez, M. C., Carrillo-González, R., Hernández Godínez, M. I. and Evangelista Lozano, S. 2017. "*Jatropha curcas* and assisted phytoremediation of a mine tailing with biochar and a mycorrhizal fungus." *International Journal of Phytoremediation* 19 (2):174-182.

González-Sandoval, M. R., Sánchez-Tovar, S. A., Márquez-Herrera, C., Lizárraga-Mendiola, L. G. and Durán-Domínguez-de-Bazúa, M. C. 2008. "Oxidación de jales ricos en pirita en un reactor a escala de banco." *Revista Latinoamericana de Recursos Naturales* 4 (2):130-138. ["Oxidation of Pyrite-Rich Tailings in a Bench-Scale Reactor." *Latin American Journal of Natural Resources*]

Gutiérrez-Bazán, J. J. 2018. *Propuesta para la disposición de jales agotados, previamente caracterizados y estabilizados, en cavidades mineras.* Master Thesis. Autonomous National University of Mexico. [*Proposal for the disposal of exhausted tailings, previously characterized and stabilized, in mining cavities.*]

Gutiérrez-Ruiz, M. E. and Moreno-Turrent, M. 2007. *Los residuos en la minería mexicana.* Instituto Nacional de Ecología y Cambio Climático. Accessed July 7, 2021. http://www2.inecc.gob.mx/publicaciones2/libros/35/los_residuos.html. [*Waste in Mexican mining.* National Institute of Ecology and Climate Change.]

Guzmán-Rangel, G. 2012. *Estrategia para la remediación de sitios contaminados con Cu, Cd, Pb y Zn aplicando tratamientos químico-agronómicos a jales minero-metalúrgicos y suelos.* Master Thesis. Autonomous National University of Mexico. [*Strategy for the remediation of sites contaminated with Cu, Cd, Pb and Zn applying chemical-agronomic treatments to mining-metallurgical tailings and soils.*]

Haddaway, N. R., Cooke, S. J., Lesser, P., Macura, B., Nilsson, A. E., Taylor, J. J. and Raito, K. 2019. "Evidence of the impacts of metal mining and the effectiveness of mining mitigation measures on social–ecological systems in Arctic and boreal regions: a systematic map protocol." *Environmental Evidence* 8 (1):1-11.

Hernández-Acosta, E., Mondragón-Romero, E., Cristobal-Acevedo, D., Rubiños-Panta, J. E. and Robledo-Santoyo, E. 2009. "Vegetación, Residuos de mina y elementos potencialmente tóxicos de un jal de pachuca, Hidalgo, México." *Revista Chapingo Serie Ciencias forestales y del Ambiente* 15 (2):109-114. ["Vegetation, mine waste and potentially toxic elements from a jal de pachuca, Hidalgo, Mexico." *Chapingo Magazine Forest and Environmental Sciences Series*]

Hernández-Ruiz-Gaytán, M. L. and Padilla-Carrillo, C. M. 2010. *Movilidad ambiental de metales en residuos mineros de la región de Zimapán, Hidalgo.* Bachelors Thesis. Autonomous National University of Mexico. [*Environmental mobility of metals in mining waste in the region of Zimapán, Hidalgo*]

Hernández-Zamora, M. A. 2009. *Estudio de la acumulación de plomo y cadmio por Asphodelus fistulosus L. y Brassica juncea L. para fitorremediar jales.* Master Thesis. Metropolitan Autonomous University. [*Study of the accumulation of lead and cadmium by Asphodelus fistulosus L. and Brassica juncea L. to phytoremediate tailings.*]

Hernández-Zamora, M. A. 2009. *Estudio de la acumulación de plomo y cadmio por Asphodelus fistulosus L. y Brassica juncea L. para fitorremediar jales.* Master Thesis. Metropolitan Autonomous University. [*Study of the accumulation of lead and cadmium by Asphodelus fistulosus L. and Brassica juncea L. to phytoremediate tailings.*]

Hudson-Edwards, K. A., Jamieson, H. E. and Lottermoser, B. G. 2011. "Mine wastes: past, present, future." *Elements* 7 (6):375-380.

Jones, H. and Boger, D. V. 2012. "Sustainability and waste management in the resource industries." *Industrial & Engineering Chemistry Research* 51 (30):10057-10065.

Juárez-Tapia, J., Patiño-Cardona, F., Roca-Vallmajor, A., Teja-Ruiz, A., Reyes-Domínguez, I., Reyes-Pérez, M., Pérez-Labra, M. and Flores-Guerrero, M. 2018. "Determination of Dissolution Rates of Ag Contained in Metallurgical and Mining Residues in the $S_2O_3^{2-}$–O_2-Cu^{2+} System: Kinetic Analysis." *Minerals* 8 (7):309.

Kossoff, D., Dubbin, W. E., Alfredsson, M., Edwards, S. J., Macklin, M. G. and Hudson-Edwards, K. A. 2014. Mine tailings dams: Characteristics, failure, environmental impacts, and remediation. *Applied Geochemistry* 51, 229-245.

Lèbre, É., Corder, G. D. and Golev, A. 2017. Sustainable practices in the management of mining waste: A focus on the mineral resource. *Minerals Engineering* 107:34-42.

Lizárraga-Mendiola, L., Ángeles-Chávez, D. E., Blanco-Piñón, A., Ramírez-Cardona, M., Olguín-Coca, F. J. and González-Sandoval, M. R. 2014. "Contamination Potential of an Urban Mine Tailings Deposit in Central Mexico- A Preliminary Estimation." *International Journal of Geosciences* 5:296-312.

Lizárraga-Mendiola, L., Durán-Domínguez, M. C. and González-Sandoval, M. R. 2008. "Environmental assessment of an active tailing pile in the State of Mexico (central Mexico)." *Research Journal of Environmental Sciences* 2 (3):197-208.

Lizárraga-Mendiola, L., González-Sandoval, M. R., Duran-Domínguez, M. C. and Márquez-Herrera, C. 2009. "Geochemical behavior of heavy metals in a Zn-Pb-Cu mining area in the State of Mexico (central Mexico)." *Environmental Monitoring and Assessment* 155 (1):355-372.

Maldonado-Villanueva, R. 2008. *Caracterización mineralógica de fases minerales metálicas en muestras de jales del distrito minero El Oro-Tlalpujahua.* Bachelors Thesis. Universidad Nacional Autónoma de México. [*Mineralogical characterization of metallic mineral phases in tailings samples from the El Oro-Tlalpujahua mining district.*]

Martínez-Soriano, C. 2011. *Evaluación de la peligrosidad de jales y lixiviados de una mina de sulfuros masivos en el Estado de México.* Bachelors Thesis. Autonomous National University of Mexico. [*Evaluation of the dangerousness of tailings and leachates from a massive sulphide mine in the State of Mexico.*]

Martínez-Soto, J. I., Teja-Ruiz, A. M., Reyes-Pérez, M., Pérez-Labra, M., Reyes-Cruz, V. E., Cobos-Murcía, J. A., Urbano-Reyes, G. and Juárez-Tapia, J. 2020. "Chemical Characterization and Mineralogical Analysis of Mining-Metallurgical Tailing from the State of Morelos." In *The Minerals, Metals & Materials Society 2020,* edited by Li, J., Zhang, M., Monteiro, S. N., Ikhmayies, S., Kalay, Y. E., Hwang, J. Y., Escobedo-Diaz, J. P., Carpenter J. S. and Brown, A. D. pp 501-510. Springer, Cham.

Molina-Garduño, A. 2018. *Geoquímica y distribución de Sb en jales del distrito minero de Zimapán, Hidalgo.* Master Thesis. Autonomous National University of Mexico. [*Geochemistry and distribution of Sb in tailings of the mining district of Zimapán, Hidalgo*]

Moreno-Tovar, R., Barbanson, L. and Coreño-Alonso, O. 2009. "Neoformación mineralógica en residuos mineros (jales) del distrito minero Zimapán, estado de Hidalgo, México." *Minería y Geología* 25 (2):1-31. ["Mineralogical neoformation in mining waste (tailings) of the Zimapán mining district, Hidalgo state, Mexico." *Mining and Geology*]

Moreno-Tovar, R., Téllez-Hernández, J. and Monroy-Fernández, M. 2012. "Influencia de los minerales de los jales en la bioaccesibilidad de arsénico, plomo, zinc y cadmio en el distrito minero Zimapán, México." *Revista Internacional de Contaminación Ambiental* 28 (3):203-2018. ["Influence of tailings minerals on the bioaccessibility of arsenic, lead, zinc and cadmium in the Zimapán mining district, Mexico." *International Journal of Environmental Pollution*]

Mussali-Galante, P., Tovar-Sánchez, E., Valverde, M., Valencia-Cuevas, L. and Rojas, E. 2013. "Evidence of population genetic effects in *Peromyscus melanophrys* chronically exposed to mine tailings in Morelos, Mexico." *Environmental Science and Pollution Research* 20:7666-7679.

Nordstrom, D. K. 2011. "Mine Waters: Acidic to Circumneutral." *Elements* 7:393-398.

Ortiz-Sánchez, O., Godelia-Canchari, S. and Soto, C. 2010. "Minería in situ. Su aplicación en un yacimiento de cobre oxidado." *Revista del Instituto de Investigaciones FIGMMG* 13 (25):31-41. ["On-site mining. Its application in an oxidized copper deposit." *Journal of the FIGMMG Research Institute*]

Park, I., Baltazar-Tabelin, C., Jeon, S., Li, X., Seno, K., Ito, M. and Hiroyoshi, N. 2019. "A review of recent strategies for acid mine drainage prevention mine tailings recycling." *Chemosphere* 219:588-606.

Patiño, F., Hernández, J., Flores, M., Reyes, I., Reyes, M. and Juárez, J. 2016. "Characterization and stoichiometry of the cyanidation reaction in NaOH of Argentian waste tailings of Pachuca, Hidalgo, México." In *Characterization of Minerals, and Materials 2016,* edited by Ikhmavies, S. J., Li, B., Carpenter J. S., Hwang, J-Y., Monteiro, S. N., Li, J., Firrao, D., Zhang, M., Peng, Z., Escobedo-Diaz, J. P. and Bai, C. pp 355-362. Springer, Cham.

Pérez-Martínez, I. 2005. *Procesos de oxidación en una presa de jales en el distrito minero de Zimapán, Hidalgo.* Master Thesis. Autonomous National University of Mexico. [*Oxidation processes in a tailings dam in the mining district of Zimapán, Hidalgo*]

Plumlee, G. and Morman, S. 2011. "Mine Wastes and Human Health." *Elements* 7:399-404.

Ramírez, E. 2001. *Almacenamiento de residuos mineros en México. Memoria del seminario Almacenamientos de Residuos Mineros CNA – SMMS - CMM, México D. F. México.* [*Storage of mining waste in Mexico. Report of the CNA - SMMS - CMM Mining Waste Storage seminar, México D. F. México.*]

Ramos-Arroyo, Y. R. and Siebe-Grabach, C. D. 2006. "Estrategia para identificar jales con potencial de riesgo ambiental en un distrito minero: estudio de caso en el Distrito de Guanajuato, México."

Revista Mexicana de Ciencias Geológicas 23 (1):54-74. ["Strategy to identify tailings with potential for environmental risk in a mining district: case study in the District of Guanajuato, Mexico." *Mexican Journal of Geological Sciences*]

Rodríguez-Valles, D. I. 2016. *Evaluación de metodologías alternas para la lixiviación de plata de jales mineros refractarios*. Master Thesis. National Polytechnic Institute. [*Evaluation of alternative methodologies for the leaching of silver from refractory mining tailings*]

Romero, F. M., Armienta, M. A. and González-Hernández, G. 2007. "Solid-phase control on the mobility of potentially toxic elements in an abandoned lead/zinc mine tailings impoundment, Taxco, Mexico." *Applied Geochemistry* 22 (1):109-127.

Romero, F. M., Armienta, M. A., Gutiérrez, M. E. and Villaseñor, G. 2008. "Factores geológicos y climáticos que determinan la peligrosidad y el impacto ambiental de jales mineros." *Revista Internacional de Contaminación Ambiental* 24 (2):43-54. ["Geological and climatic factors that determine the danger and environmental impact of mining tailings." *International Journal of Environmental Pollution*]

Sánchez-López, A., González-Chávez, M. C., Carrillo-González, R., Vangronsveld, J. and Díaz-Garduño, M. 2015. "Wild flora of Mine Tailings: Perspectives for Use in Phytoremediation of potentially toxic elements in a semi-arid region in Mexico." *International Journal of Phytoremediation* 17 (5):476-484.

Santiago-Dionisio, M. C. 2011. *Caracterización química y fitorremediación de los jales mineros de "El Fraile," de Taxco, De Alarcón, Gro., México*. Doctoral Thesis. Autonomous University of Guerrero. [*Chemical characterization and phytoremediation of "El Fraile" mining tailings, from Taxco, De Alarcón, Gro., Mexico*.]

Santos-Jallath, J. E., Coria-Camarillo, J., Huezo-Casillas, J. and Rodríguez-Cruz, G. 2013. "Influencia de jales mineros sobre el río Maconí, Querétaro, y evaluación del proceso de atenuación natural por dispersión." *Boletín de la Sociedad Geológica Mexicana* 65(3): 645-660. ["Influence of mining tailings on the Maconí River, Querétaro, and evaluation of the natural attenuation process by dispersion." *Bulletin of the Mexican Geological Society*]

Santos-Martínez, C. A. 2006. *Determinación de la peligrosidad de jales mineros con base en la normatividad ambiental mexicana y su efecto potencial en el ambiente*. Bachelors Thesis. Autonomous National University of Mexico. [*Determination of the dangerousness of mining tailings based on Mexican environmental regulations and its potential effect on the environment.*]

Santoyo-Martínez, M., Mussali-Galante, P., Hernández-Plata, I., Valencia-Cuevas, L., Flores-Morales, A., Ortiz-Hernández, L., Flores-Trujillo, K., Ramos-Quintana, F. and Tovar-Sánchez, E. 2020. "Heavy metal bioaccumulation and morphological changes in *Vachellia campechiana* (Fabaceae) reveal its potential for phytoextraction of Cr, Cu, and Pb in mine tailings." *Environmental Science and Pollution Research* 27 (10):11260-11276.

Secretaría de Economía (2021). *Minería*. Accessed July 7, 2021. https://www.gob.mx/se/acciones-y-programas/mineria. [*Mining.*]

Secretaría de Medio Ambiente y Recursos Naturales. (2015). *Informe de la Situación del Medio Ambiente en México. Edición 2015.* Secretaría del Medio Ambiente y Recursos Naturales. México. Accessed July 7, 2021. https://apps1.semarnat.gob.mx:8443/dgeia/informe15/tema/pdf/Informe15_completo.pdf. [Ministry of the Environment and Natural Resources. (2015). *Report on the Situation of the Environment in Mexico. Edition 2015.* Ministry of the Environment and Natural Resources.]

Servicio Geológico Mexicano (2018a). *Anuario estadístico de la minería mexicana, 2017.* Subsecretaría de Minería. México. Accessed July 7, 2021. http://www.sgm.gob.mx/productos/pdf/Anuario_2017_Edicion_2018.pdf. [Mexican Geological Survey (2018a). *Statistical Yearbook of Mexican Mining, 2017.* Undersecretariat of Mining.]

Servicio Geológico Mexicano. (2018b). *Panoramas Mineros Estatales.* Accessed July 7, 2021. https://www.gob.mx/sgm/articulos/consulta-los-panoramas-mineros-estatales. [*State Mining Panoramas.*]

Servicio Geológico Mexicano. (2020). *Geoinfomex*. Accessed July 7, 2021. https://www.sgm.gob.mx/GeoInfoMexGobMx/.

Sistema Integral Sobre Economía Minera. (2018). *Sistema Integral sobre Economía Minera.* Accessed July 7, 2021. https://www.sgm.gob.mx/SINEMGobMx/Producción_minera.jsp. [*Comprehensive System on Mining Economy.*]

Sistema Integral Sobre Economía Minera. (2019). *Minería de México por Entidad Federativa.* Accessed July 7, 2021. https://www.sgm.gob.mx/SINEMGobMx/mineria_mexico.jsp. [*Mining of Mexico by State.*]

Söderholm, K., Söderholm, P., Helenius, H., Pettersson, M., Viklund, R., Masloboev, V., Mingaleva, T. and Petrov, V. 2015. "Environmental regulation and competitiveness in the mining industry: Permitting processes with special focus on Finland, Sweden and Russia." *Resources Policy* 43:130-142.

Solís-Miranda, B. 2016. *Aislamiento de bacterias de jales mineros y análisis de su potencial para la remediación de sitios contaminados con metales pesados.* Master Thesis. Autonomous University of Morelos State. [*Isolation of bacteria from mining tailings and analysis of their potential for remediation of sites contaminated with heavy metals.*]

Tovar-Sánchez, E., Cervantes, L. T., Martínez, C., Rojas, E., Valverde, M., Ortiz-Hernández, M. L. and Mussali-Galante, P. 2012. "Comparision of two wild rodent species as sentinels of environmental contamination by mine tailings." *Environmental Science and Pollution Research* 19:1677-1686.

Velasco, J., De la Rosa, D., Solórzano, G. and Volke, T. 2004. *Primer informe del proyecto: Evaluación de tecnologías de remediación para suelos contaminados con metales.* Secretaría del Medio Ambiente y Recursos Naturales- Instituto Nacional de Ecología. [*First project report: Evaluation of remediation technologies for soils contaminated with metals.*]

Volke, T., Velasco, J. and De la Rosa, D. 2005. *Suelos contaminados por metales y metaloides: muestreo y alternativas para su remediación.* Secretaría del Medio Ambiente y Recursos Naturales. [*Soils contaminated by metals and metalloids: sampling and alternatives for their remediation*]

Volpi-León, V. 2017. *Efecto de corrosión en concreto reforzado elaborado con desecho minero (jal).* Master Thesis. Benemérita

Autonomous University of Puebla. [*Corrosion effect in reinforced concrete made with mining waste (jal).*]

Wang, C., Harbottle, D., Liu, Q. and Xu, Z. 2014. "Current state of fine mineral tailings treatment: A critical review on theory and practice." *Minerals Engineering* 58:113-131.

Zamora-Duarte, Y. 2008. *Efecto lixiviante causado bacterias mesófilas, autóctonas e inoculadas, sobre jales frescos de Minera Tizapa por pruebas de intemperismo acelerado.* Bachelors Thesis. Universidad Nacional Autónoma de México. [*Leaching effect caused by mesophilic, autochthonous and inoculated bacteria, on fresh tailings from Minera Tizapa by accelerated weathering tests*]

Zúñiga-Vázquez, D. V. (2014). *Evaluación de la movilidad de Cd, Pb, Zn, Fe y As bajo diferentes condiciones fisicoquímicas en jales mineros del municipio de Zimapán, Hidalgo.* Bachelors Thesis. Autonomous National University of Mexico. [*Evaluation of the mobility of Cd, Pb, Zn, Fe and As under different physicochemical conditions in mining tailings of the municipality of Zimapán, Hidalgo*]

Zúñiga-Vázquez, D. V. 2019. *Evidencias de dispersión de jales mineros sobre un transecto en suelos de la región de Zimapán, Hidalgo.* Master Thesis. Autonomous National University of Mexico. [*Evidence of dispersion of mining tailings on a transect in soils of the Zimapán region, Hidalgo*]

Biographical Sketch

Patricia Mussali Galante

Affiliation: Universidad Autónoma del Estado de Morelos.

Education: PhD in Biological Sciences.

Research and Professional Experience: Ecotoxicology.

Professional Appointments: Senior Professor-Research.

Publications from the Last 3 Years:

1. De la Cruz-Guarneros Natalia, Tovar-Sánchez Efraín y Mussali-Galante Patricia. 2021. Assessing effects of chronic heavy metal exposure through a multibiomarker approach: the case of *Liomys irroratus* (Rodentia: Heteromyidae). *Environmental Science and Pollution Research*, 1-15. https://doi.org/10.1007/s11356-021-14855-w.
2. Ortiz-Hernández Ma. Laura, Gama-Martínez Yitzel, Fernández-López Maikel, Castrejón-Godínez, María Luisa, Encarnación Sergio, Tovar-Sánchez Efraín, y Mussali-Galante Patricia. 2021. Transcriptomic analysis of *Burkholderia cenocepacia* CEIB S5-2 during methyl parathion degradation. *Environmental Science and Pollution Research*, 1-18. https://doi.org/10.1007/s11356-021-13647-6.
3. Valencia-Cuevas, Almendra Rodríguez-Domínguez, Patricia Mussali-Galante, Fernando Ramos-Quintana, Efraín Tovar-Sánchez. 2020. Influence of edaphic factors along an altitudinal gradient on a litter arthropod community in an Abies-Quercus forest in Mexico. *Acta Oecologica*. 108: DOI: 10.1016/j.actao.2020.103609.
4. Dalia A. Muro-González, Patricia Mussali-Galante, Leticia Valencia Cuevas, Karen Flores-Trujillo y Efraín Tovar-Sánchez. 2020. Morphological, physiological, and genotoxic effects of heavy metal bioaccumulation in *Prosopis laevigata* reveal its potential for phytoremediation. *Environmental Science and Pollution Research*. *27*(32), 40187-40204. https://doi.org/10.1007/s11356-020-10026-5.
5. Santoyo-Martínez, M., Mussali-Galante, P., Hernández-Plata, I., Valencia-Cuevas, L., Flores-Morales, A., Ortiz-Hernández, L., Tovar-Sánchez, E. 2020. Heavy metal bioaccumulation and morphological changes in *Vachellia campechiana* (Fabaceae) reveal its potential for phytoextraction of Cr, Cu, and Pb in mine tailings. *Environmental Science and Pollution Research*, 1-17. https://doi.org/10.1007/s11356-020-07730-7 (corresponding author).

6. Luz Breton-Deval, Ayixon Sánchez-Reyes, Alejandro Sánchez-Flores, Katy Juárez, Ilse Salinas-Peralta, and Patricia Mussali-Galante. Functional Analysis of a Polluted River Microbiome Reveals a Metabolic Potential for Bioremediation. 2020. *Microorganisms*. 8, 554.
7. Flores-Trujillo, A. K. I., Mussali-Galante, P., Calero de Hoces, M., Blázquez-García, G., Saldarriaga-Noreña, H. A., Rodríguez-Solís, A., Tovar-Sánchez, E., Sánchez-Salinas E. & Ortiz-Hernández, L. 2020. Biosorption of heavy metals on *Opuntia fuliginosa* and *Agave angustifolia* fibers for their elimination from water. *International Journal of Environmental Science and Technology*. doi 10.1007/s13762-020-02832-8.
8. Hernández-Plata I., Rodríguez VM., Tovar-Sánchez E., Carrizalez L., Villalobos P., Mendoza-Trejo MS., Mussali-Galante P. 2020. Metal brain bioaccumulation and neurobehavioral effects on the wild rodent *Liomys irroratus* inhabiting mine tailing areas. *Environmental Science and Pollution Research*. https://doi.org/10.1007/s11356-020-09451-3.
9. Salazar-Ramírez, G., Flores-Vallejo, R. D. C., Rivera-Leyva, J. C., Tovar-Sánchez, E., Sánchez-Reyes, A., Mena-Portales, J., Mussali-Galante, P. 2020. Characterization of Fungal Endophytes Isolated from the Metal Hyperaccumulator Plant *Vachellia farnesiana* Growing in Mine Tailings. *Microorganisms*, 8(2), 226. doi:10.3390/microorganisms8020226.
10. Rodríguez A., Castrejón- Godínez ML., Salazar- Bustamante E., Gama- Martínez Y., Sánchez- Salinas E., Mussali- Galante P., Tovar- Sánchez E., Ortiz- Hernández ML., 2020. Omics Approaches to Pesticide Biodegradation. *Current Microbiology* 77:545–563. doi: 10.10007/s00284-020-01916-5.
11. Mussali-Galante P. y Tovar-Sánchez. E. 2020. Diversidad genética de especies centinela como bioindicador de la salud poblacional. In: *La biodiversidad en Morelos: Estudio del Estado 2. Vol II.* Comisión Nacional para el Conocimiento de la Biodiversidad. Printed in Mexico. Pp. 430-434. ISBN: 978-607-8570-41-6. [Genetic diversity of sentinel species as a bioindicator of population health. In: *Biodiversity in Morelos: Study of the State 2. Vol II.*]

12. Castillo-Mendoza E., Mussali-Galante P., Valencia-Cuevas L., Flores Franco G., Castrejón Godínez L., Hernández-Plata I, Galván MA, Rodríguez Solís AJ and Tovar-Sánchez E. 2020. Half of Mexico's invasive plant species may be useful for heavy metal phytoremediation. In: Vinícius Londe (ed.). *Invasive species: Ecology, Impacts, and Potential Uses.* Chapter 8. Pp. 247-309. Nova Science Publishers Inc, USA. Pp. 247-309. ISBN: 978-1-53617-890-6.
13. Castrejón-Godínez, M. L., Sánchez-Salinas, E., Mussali-Galante, P., Salazar-Bustamante, E., Encarnación, S., Rodríguez, A., Tovar-Sánchez. E. 2019. Transcriptional analysis reveals the metabolic state of *Burkholderia zhejiangensis* CEIB S4-3 during methyl parathion degradation. International Biodeterioration & Biodegradation. *Peer J.* doi: 10.7717/peerj.6822.
14. López-Caamal A., Ferrufino-Acosta L. F., Díaz-Maradiaga R. F., Rodríguez-Delcid, D., Mussali-Galante P., Tovar-Sánchez E. 2019. Species distribution modelling and cpSSR reveal population history of the Neotropical annual herb *Tithonia rotundifolia* (Asteraceae). *Plant Biology.* doi: 10.1111/plb.12925 (corresponding author).
15. Tovar-Sánchez, E., Suarez-Rodríguez, R., Ramírez-Trujillo, A., Valencia-Cuevas, L., Hernández-Plata, I., and Mussali-Galante, P. 2019. The use of biosensors for biomonitoring environmental metal pollution. In: Rinken T. and Kivira K (eds), *Environmental Biosensors*, IntechOpen, London, UK. ISBN: 978-953-307-486-3. doi: http://dx.doi.org/10.5772/intechopen.84309.

In: Environmental Management
Editor: Miguel Fischer ISBN: 978-1-68507-019-9

Chapter 4

MINING WASTE MANAGEMENT ALTERNATIVES, BIOLOGICAL MONITORING AND BIOREMEDIATION STRATEGIES

María Luisa Castrejón-Godínez[1], Abigail Díaz-Armendáriz[2], Alexis Joavany Rodríguez-Solís[3], Patricia Mussali-Galante[3] and Efraín Tovar-Sánchez[3,*]

[1]Facultad de Ciencias Biológicas,
Universidad Autónoma del Estado de Morelos,
Cuernavaca, Morelos, México
[2]Especialidad en Gestión Integral de Residuos,
Universidad Autónoma del Estado de Morelos,
Cuernavaca, Morelos, México
[3]Centro de Investigación en Biotecnología,
Universidad Autónoma del Estado de Morelos,
Cuernavaca, Morelos, México

[*] Corresponding Author's E-mail: efrain_tovar@uaem.mx.

Abstract

Mining is one of the most important extractive activities worldwide. It has a significant impact on the economic development of different regions in many countries and supplies the growing demand for mineral resources for various industries. However, this economic activity is well recognized as an environmental hazard due to the release of several pollutants to nature. Among the most relevant environmental concerns related to mining activities are solid and liquid waste generation, including mine tailings and acidic drainage. Adequate management of such waste is an important task for avoiding negative impacts on the environment. Biomonitoring strategies in or near sites impacted by mining activities give information on the overall environmental health, enabling the identification of bioaccumulation and biomagnification processes, and opportunities for remediation of the mining-impacted sites and implementation of suitable waste management strategies. In the present chapter, we review the biomonitoring strategies employed to evaluate the negative environmental impact of mining waste, the remediation alternatives for mining-polluted sites and environmental management approaches.

Keywords: mining waste, management, bioindicators, heavy metal, bioremediation

Introduction

Worldwide, mining is one of the oldest and most essential activities of humanity, contributing to the economic growth of different regions (Rebello et al. 2021). In several countries, mining is one of the most important primary economic activities. This industry carries out the exploration, exploitation, smelting and refining of different minerals extracted from nature (Aznar-Sánchez et al. 2018). However, although it has been of great importance, it has generated social, economic, environmental and human health impacts related to the pollution of ecosystems, due to the generation of mining wastes called mine tailings. The mine tailings contain non-biodegradable and potentially toxic pollutants such as heavy metals (Yadav et al. 2017; Ortiz-Hernández et al. 2018).

Due to this problem it is essential to look for alternatives that can reduce the negative effects on the environment. Among these alternatives are physical and chemical treatments but these methods generate waste that must be disposed of, increasing its cost and reducing its effectiveness. On the other hand, biological treatment is based on the use of bacteria, fungi, plants or enzymes derived from these organisms. It is a profitable and environmentally friendly technique, and their purpose is to degrade, transform to innocuous metabolic products, or remove contaminants. This kind of treatment is called bioremediation and uses techniques such as biosorption, phytoremediation, bioaccumulation, bioprecipitation, bioleaching, among others (Le Cloirec and Andrès 2005; Volke et al. 2005; Lakherwal 2014; Ayangbenro and Babalola 2017; Ayangbenro et al. 2018). In this chapter, we will address the management of mining waste, the impacts of heavy metals on the environment and human health, biological monitoring, as well as the most widely used bioremediation strategies to remediate sites contaminated with heavy metals.

Mining Waste

The mining industry generates large amounts of wastes. It has been reported that this industry generates about 65 billion tons of waste per year, of which 51 billion tons are rock waste and 14 billion tons are mining waste with a particle size smaller than 120 μm, denominated mine tailings (Jones and Boger 2012; Ortiz-Hernández et al. 2018). Mining waste is a complex mixture of rock, small particle mine tailings (which contain heavy metals), as well as water and chemical employed in the process (Lèbre et al. 2016).

Heavy metals are chemical elements that show densities higher than 5 g/cm^3. They are found in the periodic table with an atomic number greater than 20 (Koller and Saleh 2018); they are not biodegradable; they accumulate in ecosystems and are biomagnified in trophic chains, causing different impacts on health and the environment. Due to this, mining waste rich in heavy metals is considered hazardous.

IMPACTS OF HEAVY METALS ON THE ENVIRONMENT AND HUMAN HEALTH

The release of heavy metals into the environment negatively impacts air, water, soil, wildlife, among others (Ortiz-Hernández et al. 2018). Heavy metals are absorbed from the soil, sediments and water, constituting a threat to different forms of life such as microorganisms, plants, animals and humans (Fryzova et al. 2017) at both the molecular and cellular levels (Geremias et al. 2012). For example, heavy metals such as Cd, Cr, Hg and Pb are found in the air, causing adverse impacts on the environment and human health. Heavy metals reach the air due to emission sources such as construction, agriculture, mining, transportation and power plants. On the other hand, heavy metals can reach water bodies by leaching or improper disposal of mining waste. These contaminants can persist in water and sediments, and they bioaccumulate in organisms causing high toxicity (Sevcikova et al. 2011). Soil is a fundamental natural resource for the organisms that inhabit it but, once mining waste is deposited in soil, heavy metals accumulate and disturb ecosystems. For example, they cause contamination of surface and underground water, entering the network trophic and generating negative impacts on the health of organisms until death. In the case of plants, they respond to the exposure of high concentrations of heavy metals in the soil through absorption and accumulation processes. Heavy metals can be taken from the soil solution by plants, mobilizing them to the roots, where the plants can absorb them.

When heavy metals are absorbed, then they accumulate in the roots of plants. Thus, the first organ affected by soil contamination is the root system, then heavy metals can become translocated to other parts of the plant as stems, leaves, fruits, and seeds (Dinu et al. 2020). Once inside the plant, these elements cause a decrease or inhibition in growth, decreasing germination, affecting stomatal function, altering photosynthesis, decreasing root, shoot and biomass growth, among others (Ayangbenro and Babalola 2017; Yadav et al. 2017).

In addition, fauna (primary consumers) absorbs heavy metals both by diffusion and by ingestion of food. There is biomagnification of these elements in the food chain through the consumption by predators, causing negative impacts on ecosystem health (Kumar et al. 2019). Heavy metals are stored in the bones and intracellular matrices of organisms while they are excreted through feces and eggs. According to Adham et al. (2011), heavy metals accumulate in the kidneys and liver of vertebrates, causing impacts on the health of organisms. Likewise, heavy metals usually enter human beings through inhalation or ingestion and once they are incorporated, they are distributed in tissues and organs. They tend to accumulate in places like the bones, liver, and kidneys over years or decades (Stankovic and Stankovic 2013).

Although some elements are essential such as Potassium (K), Magnesium (Mg), Cobalt (Co), Molybdenum (Mb), Manganese (Mn), others are toxic even in small concentrations and are not essential for biological functions, causing effects on human health (Amadi et al. 2020). Among these non-essential heavy metals is Lead (Pb) which causes in humans; anorexia, neuronal damage, infertility, kidney damage, short-term memory loss, fatigue, irritability, hypertension, cancer, coordination problems, cardiovascular diseases, among others (Ayangbenro and Babalola 2017; Yadav et al. 2017). For its part, Arsenic (As) also known as the "slow killer," causes skin lesions, hearing loss, respiratory, neurological, cardiovascular and immune complications, as well as effects on the hematopoietic and endocrine system, diabetes, liver and kidney dysfunction (Abdul et al. 2015; Prakash and Verma 2021). It also affects cellular processes such as oxidative phosphorylation and ATP synthesis (Jomova et al. 2011; Prakash et al. 2016).

Cadmium (Cd) causes kidney disorders, anemia, anosmia, cardiovascular diseases, hypertension, Itai-Itai disease, bone pain, osteoporosis, among others (Genchi et al. 2020; Nkwunonwo et al. 2020). Mercury (Hg) for its part causes death, delay, dysarthria, blindness, neurological deficits, hearing loss, abdominal pain, indigestion, ulcers, bloody diarrhea, destruction of intestinal flora, kidney damage, alteration in production and function of immune cells, sclerosis, arthritis, thyroiditis, epilepsy, psoriasis, schizophrenia, scleroderma, infertility, effects on the endocrine, nervous and reproductive systems (Rice et al. 2014). For its part, Chromium (Cr) is a necessary micronutrient for metabolism in humans and some animals. However, concentrations above 50g cause skin damage and cancer (Stankovic and Stankovic 2013).

MINING WASTE MANAGEMENT ALTERNATIVES

Due to the environmental and health problems caused by mining waste, different sustainable alternatives have been proposed for its adequate management. For example, Lèbre and Corder (2015) and Lèbre et al. (2016) proposed a hierarchy of mining waste management where reducing the generation of waste and the recovery of minerals is prioritized, including reprocessing, that refers to the extraction of the more valuable resources that remain in the waste, using technologies such as a bioreactor, phytomining, filtration, reverse osmosis, as well as ion exchange resins (Lottermoser 2011).

For its part, downcycling is defined as using all the material that is discarded. For example, tailings can be used as road construction material, the manufacture of blocks and bricks, glass production, as raw material for the production of cement and concrete, among others (Lottermoser 2011; Edraki et al. 2014), which reduces energy consumption, greenhouse gas emissions and waste generation. Each of the aforementioned stages can be combined to optimize economic gains for the mining industries but at the same time mitigate environmental impacts, guarantee community safety, human well-being, the rights of the inhabitants and the quality of life of current and future generations (Gomes et al. 2014).

Lottermoser (2011) points out several reuse and recycling options for mining waste. For example, rock waste can be used as material for filling holes, as aggregate in roads and building construction, asphalt component and raw material for cement and concrete. In the case of mine water, it can be used in industry and the agricultural sector, it can be a cooling or heating agent. The mine-drainage sludge can be used as an adsorbent to remove phosphate from wastewater. As mentioned above, in the case of mine tailings, they can be used for the manufacture of bricks, cement, as sources of alum, those that are rich in Cu can be used as thinners for paints, among other options. Finally, metallurgical waste can be used as a raw material for glass, tiles, cements, ceramics, and bricks.

Dold (2008) points out that exploring of a mineral deposit should focus on the search for "clean minerals," starting the exploitation of minerals it is essential to carry out an environmental impact study detailing the natural conditions of the area before mining activities.The management of mining waste has become a field of research thanks to social awareness of the environmental problems generated by mining and the growing demand for more sustainable production methodologies. The lines of research are oriented to the application of bioremediation through the use of microorganisms, algae and plants (Aznar-Sánchez et al. 2018).

Bioindicators of Heavy Metal Pollution

However, there are sites contaminated with mining waste, so sentinel species or biological indicators are used to evaluate the environmental impacts of such wastes. Sentinel species enables the evaluation of the environmental impact within ecosystems and the possible effects on the health of the populations by the exposure to heavy metals (Amadi et al. 2020). According to Beeby (2001), the selection criteria that organisms must meet to be considered sentinel species, in addition to being capable of accumulating high concentrations of pollutants in their tissues (Stankovic and Stankovic 2013), is that they must be ubiquitous species, that is, they inhabit environments both in the presence and absence of pollutants. In addition, sedentary lifestyle, simple sampling, classification and storage, medium to long generation rate, easy taxonomy, and easily cultivable in the laboratory (Gerhardt 2002; Berthet 2012).

Within the sentinel species employed to evaluate the environmental impact of heavy metals, microorganisms have been reported, such as bacteria, protozoa and algae (Stankovic and Stankovic 2013), plants

(Kord et al. 2010; Amadi et al. 2020), bats (Zukal et al. 2015), fish (Authman et al. 2015), small mammals such as rodents (Espinosa-Reyes et al. 2010; Tovar-Sánchez et al. 2012; Hernández-Plata et al. 2020), birds (Amadi et al. 2020), cats (Aeluro and Kavanagh 2021; Esposito et al. 2019; Amadi et al. 2020), dogs (Esposito et al. 2019; Amadi et al. 2020), nematodes (Baruš et al. 2007; Šalamún et al. 2012), mollusks (Amadi et al. 2020), among others.

Heavy Metal Bioremediation Strategies

The use of microorganisms stands out among the most used strategies for the bioremediation of sites contaminated with heavy metals. These microorganisms chemically transform or reduce some heavy metals through their metabolism. For example, there are bacteria such as *Pseudomonas fluorescens* LB300, *Bacillus* sp., *Enterobacter cloacae* HO1 and *Enterobacter aerogenes* (Thacker et al. 2006) that have the ability to reduce hexavalent chromium Cr (VI), which is toxic and mutagenic, to its trivalent form Cr (III) which is less toxic and an essential micronutrient for many microorganisms (Garbisu and Alkorta 2003).

Wang and Shen (1995) report different bacteria such as *Achromobacter eurydice, Aeromonas dechromatica, Agrobacterium radiobacter, Bacillus cereus, Bacillus subtilis, Desulfovibrio vulgaris, Escherichia coli, micrococcus roseus, Pseudomonas aeruginosa, Pseudomonas dechromaticans, Pseudomonas chromatophila, Pseudomonas ambigua* G-1 and *Pseudomonas putida* PRS 2000, can transform Cr (VI) to Cr (III). This chromium reduction can be both aerobic and anaerobic. Aerobic activity is generally associated with reductases using NADH as an electron donor, while anaerobic through soluble reductase, membrane-bound reductase, or by both reductases. In addition, microorganisms are capable of enzymatically reducing other heavy metals such as mercury (Hg), for example, *Shewanella oneidensis* MR-1, *Pseudomonas*, *Escherichia coli* and *Staphylococcus aureus*, which reduce ionic mercury Hg (II) to elemental mercury Hg (0). The latter is volatile at room temperature and can reduce it through a resistance system based on a cluster of genes (*mer* operon) located in plasmids and chromosomes that encode for transporters (*mer*T, *mer*P and/or *mer*C, *mer*F), regulators (*mer*R) and a mercury reductase enzyme (*mer*A) (Nascimento and Chartone-Souza 2003; Wiatrowski et al. 2006).

Another strategy for the removal of heavy metals is biosorption which includes absorption (where the particles are inside the material), adsorption (the particles adhere to the surface), ion exchange (heavy metals are exchanged with the counter ions of the polysaccharides of the cell walls of the microorganisms), complexation on the surface (formation of complexes on the cell surface after the interaction between the metal and the amino and carboxyl groups of the polysaccharides of the cell wall, which facilitates their solubilization and leaching of metals, favoring biosorption) and precipitation (microorganisms respond to the presence of metals, which favors the precipitation process) (Ahalya et al. 2003; Masindi and Muedi 2018). Biosorption uses biological materials called biosorbents, which are inexpensive and abundant in nature (Mustapha et al. 2015). Table 1 shows the types of biosorbents used by different authors, including microorganisms (algae, bacteria, fungi, yeasts) and agro-industrial waste. The table also shows the removal percentages of the metals from the investigations.

Bacteria have developed resistance to heavy metals and detoxification mechanisms which can be intracellular and extracellular sequestration, exporting metals out of the cell and reducing the permeability of the membrane that has allowed them to remove these pollutants from the environment. On the other hand, reports showed that algae absorb around 15 to 84% of metals. Brown marine algae can absorb metals such as Cd, Ni and Pb through chemical groups found on cell surfaces such as carboxyl, sulfonate, amino, and sulfhydryl. In addition, fungi have also been reported as a material for heavy metal biosorption (Mustapha et al. 2015).

Phytoremediation is the use of plants to accumulate, absorb, transform, mineralize or stabilize pollutants such as the heavy metals present in the environment (Núñez-López et al. 2008; Delgadillo-López et al. 2011). This applied remediation strategy permits reducing heavy metal concentrations in polluted sites through *in situ* or *ex situ* approaches (Ghosh and Singh 2005). Different phytoremediation techniques have been reported, including phytoextraction, also known as phytoaccumulation, phytobsorption or phytosecustration, denominated as the absorption of the pollutants present in the soil, in water or sediments by the roots of the plants and their translocation to the aerial biomass. In phytofiltration the pollutants are absorbed or adsorbed by the roots of the plants, known as rhizofiltration, by seedlings called blastofiltration, and by the use of plant shoots called caulofiltration.

Phytostabilization is also called phyto-immobilization. This technique reduces the mobility of pollutants through the use of plants, thereby avoiding the migration of pollutants to the environment. Phytovolatilization consists of the absorption of pollutants present in the environment by plants, transforming them into volatile elements and subsequently releasing them into the atmosphere through their leaves. Finally, phytodegradation, consists of the degradation of absorbed pollutants by the activity of different enzymes produced by plants, such as dehalogenases, oxygenases and reductases. In the case of heavy metals, plants can carry out biosorption and can be transformed but not degraded (Ghosh and Singh 2005; Jadia and Fulekar 2009; Ali et al. 2013).

Different authors have reported plant species that can accumulate heavy metals. These plants are called hyperaccumulators and according to the global hyperaccumulators database (GHD; http://hyperaccumulators.smi.uq.edu.au/collection/), currently, 721 plant species have been reported as hyperaccumulators. They belong to the following plant groups: two from Commelinids, 701 from Eudicots, one from Magnoliids, nine from Monocots and eight from Pteridophyta (Mota-Merlo and Martos 2021). Furthermore, Reeves et al. (2018) report 532 species of nickel hyperaccumulator plants, 53 for copper, 42 for cobalt, 42 for manganese, 41 for selenium, 20 for zinc, eight for lead, seven for cadmium, five for arsenic, two for rare earth elements and two for thallium. Shrivastava et al. (2019) point out that these species belong to 52 families and 130 genera and that the most representative families are the Brassicaceae with 83 species and the Phyllanthaceae with 59 species.

Table 1. Types of biosorbents used for the removal of heavy metals

Biosorbent	Species	Heavy metal	Remotion (%)	Reference
Algae	*Spirogyra hyalina*	Hg	89.0	Kumar and Oommen 2012
		Pb	83.0	
	Cladophora albida	Cd	98.9	Deng et al. 2009
		Cu	97.5	
		Pb	98.1	
		Zn	100	
	Pithophora odeogonia	Pb	97.0	Singh et al. 2007
	Spirogyra neglecta	Pb	89.0	
	Sargassum muticum	Cd	95.0	Lodeiro et al. 2004
Bacteria	*Bacillus cereus S5*	Cd	>80.0	Wu et al. 2016
	Bacillus sp.	Cd	72.9	Azzam and Tawfik 2015
		Cu	88.8	
		Pb	99.5	
	Pseudomonas putida	Cd	93.0	Nanganuru and Korrapati 2012
	Penicillium chrysogenum	Al	86.8	Ahmad et al. 2011
		Fe	97.8	
		Mn	77.6	
		Ni	57.3	
	Actinomycetes sp.	Cd	>95.0	Kefala et al. 1999

Table 1. (Continued)

Biosorbent	Species	Heavy metal	Remotion (%)	Reference
Fungi	*Pleurotus ostreatus*	Cd	45.9-61.1	Yang et al. 2017
		Cr	29.4-64.5	
		Pb	99.9-100	
	Trametes versicolor	Cu	65.4	Şahan et al. 2010
	Agaricus macrospores	Cu	96.0	Melgar et al. 2007
		Pb	89.0	
	Rhizopus aarhizus	Cu	95.5	Preetha and Viruthagiri 2007
		Cr	93.8	
		Ni	61.4	
	Aspergillus niger	Cr	99.2	Bai and Abraham 2001
Yeast	*Saccharomyces cerevisiae*	Ni	96.0	Machado et al. 2008
Agroindustrial waste	*Agave angustifolia*	Cd	71.0	Medina-Acevedo 2020
		Pb	80.0	
	Opuntia sp.	Zn	97-99	Flores-Trujillo 2016

To consider a plant species as hyper-accumulating it must tolerate high levels of heavy metals both in the roots and in the shoots, transfer the chemical element from the roots to the shoots and must have a rapid absorption rate of the metal (Muszynska and Hanus-Fajerska 2015). In addition, it must present in its foliar tissue in dry weight >100 µg/g cadmium, thallium or selenium; >300 µg/g of cobalt, copper or chromium; >1,000 µg/g of nickel, arsenic, lead and >10,000 µg/g of zinc and manganese when plants are grown in their natural habitat (Reeves et al. 2018; Shrivastava et al. 2019).

The potential of heavy metal hyperaccumulator plants allows them to be used as an alternative to remove large amounts of heavy metals present in the soil. However, it must be noted that some species are not native, so there is a risk of introducing invasive species to new habitats (Mota-Merlo and Martos 2021). In addition, taking into account collecting the plant biomass after the phytoremediation process, since the elements are persistent and can be released into the environment by leaching and other natural processes, causing contamination of the soil, surface and groundwater, generally threatening the life of ecosystems. According to Núñez-López et al. (2008), biomass can be confined in sanitary landfills and used as compost. However, these alternatives may not be viable because heavy metals can return to the environment.

An alternative is phytomining, wich is an *in situ* biological treatment where the plants accumulate heavy metals, either naturally or induced to do so by soil amendments. Later the plants are harvested, left to dry, compact and incinerated to recover valuable metals from the ash, such as Nickel (Ni) and gold (Au) (Anderson et al. 2003; Heilmeier 2021). Another alternative is the stabilization/solidification technique, which is a successful strategy for the treatment of soils contaminated with heavy metals in which Portland cement is used as a binder for the retention of heavy metals such as Cr, Cu and Pb (Pollettini et al. 2004; Al-Kimdi 2019). The ashes from the plant mining operation can be used for stabilization/solidification. Another alternative is that after incinerating the dry biomass of the plants used in the phytoremediation process, they can be taken to final disposal by some of the companies authorized to handle hazardous waste.

On the other hand, these plants are of great importance to continue with the research related to OMICS approaches. They allow us to learn more about the mechanisms of plants and through genetic engineering to obtain more efficient plants for phytoremediation.

Conclusion

The mining industry negatively impacts the environment by generating and releasing different waste, so alternatives are being sought for the proper management of mining waste and the remediation of contaminated sites. Different alternatives have been proposed to prevent, minimize, reuse and recycle mining waste, mainly in developed countries. However, in poor and developing countries, waste does not receive adequate management or treatment, which generates serious environmental contamination problems and creates risks to human health in the populations surrounding the mining facilities; due to this, the implementation of sustainable management strategies for handling this waste is of such relevance. Furthermore, the implementation of remediation strategies for those places with abandoned mining waste derived from past mining activities, and therefore constituting an environmental liability, need to recover the elements that are still in the waste, as well as mitigating the environmental impact on ecosystems and guaranteeing the quality of life of current and future generations. Among the multiple strategies proposed, bioremediation through the use of microorganisms and plants stands out as an efficient, low-cost and environmentally friendly alternative. However, it is necessary to continue with the investigations of innovative alternatives to eliminate this type of waste.

References

Abdul, K. S. M., Jayasinghe, S. S., Chandana, E. P., Jayasumana, C., and De Silva, P. M. C. 2015. "Arsenic and human health effects: A review." *Environmental Toxicology and Pharmacology 40*(3): 828-846.

Adham, K. G., Al-Eisa, N. A., and Farhood, M. H. 2011. "Risk assessment of heavy metal contamination in soil and wild Libyan jird Meriones libycus in Riyadh, Saudi Arabia." *Journal of Environmental Biology* 32(6): 813-819.

Aeluro, S., and Kavanagh, T. J. 2021. "Domestic cats as environmental lead sentinels in low-income populations: a one health pilot study sampling the fur of animals presented to a high-volume spay/neuter clinic." *Environmental Science and Pollution Research* 1-15.

Ahalya, N., Ramachandra, T. V., and Kanamadi, R. D. 2003. "Biosorption of heavy metals." *Research Journal of Chemistry Environment* 7(4): 71-79.

Ahmad, B., Bhatti, H. N., and Ilyas, S. 2011. "Bio-extraction of metal ions from laterite ore by *Penicillium chrysogenum*." *African Journal of Biotechnology* 10(54): 11196-11205.

Ali, H., Khan, E., and Sajad, M. A. 2013. "Phytoremediation of heavy metals-concepts and applications." *Chemosphere* 91(7): 869-881.

Al-Kindi, G. Y. 2019. "Evaluation the solidification/stabilization of heavy metals by Portland cement." *Journal of Ecological Engineering* 20(3): 91-100.

Amadi, C. N., Frazzoli, C., and Orisakwe, O. E. 2020. "Sentinel species for biomonitoring and biosurveillance of environmental heavy metals in Nigeria." *Journal of Environmental Science and Health, Part C* 38(1): 21-60.

Anderson, C. W. N., Stewart, R. B., Moreno, F. N., Gardea-Torresdey, J. L., Robinson, B. H., and Meech, J. 2003. "Gold phytomining. Novel Developments in a Plant-based Mining System." In: *Proceedings of the Gold 2003 Conference: New Industrial Applications of Gold.* pp. 35–45.

Authman, M. M., Zaki, M. S., Khallaf, E. A., and Abbas, H. H. 2015. "Use of fish as bio-indicator of the effects of heavy metals pollution." *Journal of Aquaculture Research & Development* 6(4): 1-13.

Ayangbenro, A. S., and Babalola, O. O. 2017. "A new strategy for heavy metal polluted environments: a review of microbial biosorbents." *International Journal of Environmental Research and Public Health* 14(1): 1-16.

Ayangbenro, A. S., Olanrewaju, O. S., and Babalola, O. O. 2018. "Sulfate-reducing bacteria as an effective tool for sustainable acid mine bioremediation." *Frontiers in Microbiology* 9: 1-10.

Aznar-Sánchez, J. A., García-Gómez, J. J., Velasco-Muñoz, J. F., and Carretero-Gómez, A. 2018. "Mining waste and its sustainable management: Advances in worldwide research." *Minerals* 8(7): 1-27.

Azzam, A. M., and Tawfik, A. 2015. "Removal of heavy metals using bacterial bio-flocculants of Bacillus sp. and Pseudomonas sp." *Journal of Environmental Engineering and Landscape Management* 23(4): 288-294.

Bai, R. S., and Abraham, T. E. 2001. "Biosorption of Cr (VI) from aqueous solution by Rhizopus nigricans." *Bioresource Technology* 79(1): 73-81.

Baruš, V., Jarkovský, J., and Prokeš, M. 2007. "Philometra ovata (Nematoda: Philometroidea): a potential sentinel species of heavy metal accumulation." *Parasitology Research* 100(5): 929-933.

Beeby, A. 2001. "What do sentinels stand for?." *Environmental Pollution* 112(2): 285-298.

Berthet, B. 2012. "Sentinel species. In *Ecological biomarkers: indicators of ecotoxicological effects.*" Amiard-Triquet, C., Amiard, J. C., & Rainbow, P. S. (Eds.). In: *Ecological biomarkers: indicators of ecotoxicological effects.* pp. 155-181. CRC Press.

Delgadillo-López, A. E., González-Ramírez, C. A., Prieto-García, F., Villagómez-Ibarra, J. R., and Acevedo-Sandoval, O. 2011. "Fitorremediación: una alternativa para eliminar la contaminación." *Tropical and Subtropical Agroecosystems* 14(2): 597-612.

Deng, L., Zhang, Y., Qin, J., Wang, X., and Zhu, X. 2009. "Biosorption of Cr (VI) from aqueous solutions by nonliving green algae Cladophora albida." *Minerals Engineering* 22(4): 372-377.

Dinu, C., Vasile, G. G., Buleandra, M., Popa, D. E., Gheorghe, S., and Ungureanu, E. M. 2020. "Translocation and accumulation of heavy metals in *Ocimum basilicum* L. plants grown in a mining-contaminated soil." *Journal of Soils and Sediments* 20(4): 2141-2154.

Dold, B. 2008. "Sustainability in metal mining: from exploration, over processing to mine waste management." *Reviews in Environmental Science and Bio/technology* 7(4): 275-285.

Edraki, M., Baumgartl, T., Manlapig, E., Bradshaw, D., Franks, D. M., and Moran, C. J. 2014. "Designing mine tailings for better environmental, social and economic outcomes: a review of alternative approaches." *Journal of Cleaner Production* 84: 411-420.

Espinosa-Reyes, G., Torres-Dosal, A., Ilizaliturri, C., Gonzalez-Mille, D., Diaz-Barriga, F., and Mejia-Saavedra, J. 2010. "Wild rodents (Dipodomys merriami) used as biomonitors in contaminated mining sites." *Journal of Environmental Science and Health, Part A* 45(1): 82-89.

Esposito, M., De Roma, A., Maglio, P., Sansone, D., Picazio, G., Bianco, R., Martinis, C., Rosao, G., Baldi, L., and Gallo, P. 2019. "Heavy metals in organs of stray dogs and cats from the city of Naples and its surroundings (Southern Italy)." *Environmental Science and Pollution Research* 26(4): 3473-3478.

Flores-Trujillo, A. K. 2016. *Remoción de metales pesados del agua a través de procesos de biosorción.* Master thesis. Autonomous University of Morelos State. Morelos, pp. 90. [*Removal of heavy metals from water through biosorption processes.*]

Fryzova, R., Pohanka, M., Martinkova, P., Cihlarova, H., Brtnicky, M., Hladky, J., and Kynicky, J. 2017. "Oxidative stress and heavy metals in plants." *Reviews of Environmental Contamination and Toxicology* 245: 129-156.

Genchi, G., Sinicropi, M. S., Lauria, G., Carocci, A., and Catalano, A. 2020. "The effects of cadmium toxicity." *International Journal of Environmental Research and Public Health* 17(11): 1-24.

Geremias, R., Bortolotto, T., Wilhelm-Filho, D., Pedrosa, R. C., and de Fávere, V. T. 2012. "Efficacy assessment of acid mine drainage treatment with coal mining waste using *Allium cepa* L. as a bioindicator." *Ecotoxicology and Environmental Safety* 79: 116-121.

Gerhardt, A. 2002. "Bioindicator species and their use in biomonitoring." *Environmental Monitoring* 1: 77-123.

Ghosh, M., and Singh, S. P. 2005. "A review on phytoremediation of heavy metals and utilization of it's by products." *Asian Journal on Energy and Environment* 6(4): 214-231.

Gomes, C. M., Kneipp, J. M., Kruglianskas, I., da Rosa, L. A. B., and Bichueti, R. S. 2014. "Management for sustainability in companies of the mining sector: an analysis of the main factors related with the business performance." *Journal of Cleaner Production* 84: 84-93.

Heilmeier, H. 2021. "Phytomining applied for postmining sites." Erickson, L. E., and Pidlisnyuk, V. (Eds.). In: *Phytotechnology with Biomass Production.* pp. 61-75. CRC Press.

Hernández-Plata, I., Rodríguez, V. M., Tovar-Sánchez, E., Carrizalez, L., Villalobos, P., Mendoza-Trejo, M. S., and Mussali-Galante, P. 2020. "Metal brain bioaccumulation and neurobehavioral effects on the wild rodent *Liomys irroratus* inhabiting mine tailing areas." *Environmental Science and Pollution Research* 27(29): 36330-36349.

Jadia, C. D., and Fulekar, M. H. 2009. "Phytoremediation of heavy metals: recent techniques." *African Journal of Biotechnology* 8(6): 921-928.

Jomova, K., Jenisova, Z., Feszterova, M., Baros, S., Liska, J., Hudecova, D., Rhodes, C. J., and Valko, M. 2011. "Arsenic: toxicity, oxidative stress and human disease." *Journal of Applied Toxicology* 31(2): 95-107.

Kefala, M. I., Zouboulis, A. I., and Matis, K. A. 1999. "Biosorption of cadmium ions by Actinomycetes and separation by flotation." *Environmental Pollution* 104(2): 283-293.

Koller, M., and Saleh, H. M. 2018. "Introductory chapter: introducing heavy metals." *Heavy Metals* 1: 3-11.

Kord, B., Mataji, A., and Babaie, S. 2010. "Pine (*Pinus Eldarica* Medw.) needles as indicator for heavy metals pollution." *International Journal of Environmental Science & Technology* 7(1): 79-84.

Kumar, J. N., and Oommen, C. 2012. "Removal of heavy metals by biosorption using freshwater alga *Spirogyra hyalina*." *Journal of Environmental Biology* 33(1): 27-31.

Kumar, S., Prasad, S., Yadav, K. K., Shrivastava, M., Gupta, N., Nagar, S., Bach, Q. V., Kamyab, H., Khan, S. A., Yadav, S., and Malav, L. C. 2019. "Hazardous heavy metals contamination of vegetables and food chain: Role of sustainable remediation approaches-A review." *Environmental Research 179*: 1-12.

Lakherwal, D. 2014. "Adsorption of heavy metals: a review." *International Journal of Environmental Research and Development* 4(1): 41-48.

Le Cloirec, P., and Andrès, Y., 2005. "Bioremediation of heavy metals using microorganisms." *Bioremediation of Aquatic and Terrestrial Ecosystems* 97-140.

Lèbre, É. and Corder, G. 2015. "Integrating industrial ecology thinking into the management of mining waste." *Resources* 4(4): 765-786.

Lèbre, É., Corder, G. D., and Golev, A. 2016. "Sustainable practices in the management of mining waste: A focus on the mineral resource." *Minerals Engineering* 107: 34-42.

Lodeiro, P., Cordero, B., Grille, Z., Herrero, R., and Sastre de Vicente, M. E. 2004. "Physicochemical studies of cadmium (II) biosorption by the invasive alga in Europe, *Sargassum muticum*." *Biotechnology and Bioengineering* 88(2): 237-247.

Lottermoser, B. G. 2011. "Recycling, reuse and rehabilitation of mine wastes." *Elements* 7(6): 405-410.

Machado, M. D., Santos, M. S., Gouveia, C., Soares, H. M., and Soares, E. V. 2008. "Removal of heavy metals using a brewer's yeast strain of *Saccharomyces cerevisiae*: the flocculation as a separation process." *Bioresource Technology* 99(7): 2107-2115.

Masindi, V., and Muedi, K. L. 2018. "Environmental contamination by heavy metals." *Heavy Metals* 10: 115-132.

Medina-Acevedo, V. M. 2020. *Remoción de plomo y cadmio mediante fibras de Agave angustifolia y un consorcio bacteriano como una alternativa de biorremediación.* Degree thesis Enviromental Science. Autonomous University of Morelos State. Morelos, pp. 68. [*Lead and cadmium removal using Agave angustifolia fibers and a bacterial consortium as a bioremediation alternative.*]

Melgar, M. J., Alonso, J., and García, M. A. 2007. "Removal of toxic metals from aqueous solutions by fungal biomass of *Agaricus macrosporus*." *Science of the Total Environment* 385(1-3): 12-19.

Mota-Merlo, M., and Martos, V. 2021. "Use of machine learning to establish limits in the classification of hyperaccumulator plants growing on serpentine, gypsum and dolomite soils." *Mediterranean Botany* 42: 1-14.

Mustapha, M. U., and Halimoon, N. 2015. "Microorganisms and biosorption of heavy metals in the environment: a review paper." *Journal of Microbial and Biochemical Technology* 7(5): 253-256.

Muszynska, E., and Hanus-Fajerska, E. 2015. "Why are heavy metal hyperaccumulating plants so amazing? *BioTechnologia." Journal of Biotechnology Computational Biology and Bionanotechnology* 96(4): 265-271.

Nanganuru, H. Y., and Korrapati, N. 2012. "Studies on biosorption of cadmium by *Pseudomonas putida*." *International Journal Engineering Research and Applications* 2(3): 2217-2219.

Nascimento, A. M., and Chartone-Souza, E. 2003. "Operon mer: bacterial resistance to mercury and potential for bioremediation of contaminated environments." *Genetics and Molecular Research* 2(1): 92-101.

Nkwunonwo, U. C., Odika, P. O., and Onyia, N. I. 2020. "A review of the health implications of heavy metals in food chain in Nigeria." *The Scientific World Journal* 1-11.

Núñez-López, R. A., Meas, Y., Gama, S. C., Borges, R. O., and Olguín, E. J. 2008. "Leaching of lead by ammonium salts and EDTA from *Salvinia minima* biomass produced during aquatic phytoremediation." *Journal of Hazardous Materials* 154(1-3): 623-632.

Ortiz Hernández, M L., Mussali Galante, P., Sánchez Salinas, E., and Tovar Sánchez, E. 2018. "Mining and mine tailings: Characterization, impacts, ecology and bioremediation strategies." Fuentes, M. S., Colin, V. L., & Saez, J. M. (Eds.). In: *Strategies for bioremediation of organic and inorganic pollutants*. pp. 190-214. CRC Press.

Polletini, A., Pomi, R., and Valente, M. 2004. "Remediation of heavy metal-contaminated soil by means of agglomeration." *Journal of Environmental Science and Health* A39: 999-1010.

Prakash, C., Soni, M., and Kumar, V. 2016. "Mitochondrial oxidative stress and dysfunction in arsenic neurotoxicity: a review." *Journal of Applied Toxicology* 36(2): 179-188.

Prakash, S., and Verma, A. K. 2021. "Arsenic: It's Toxicity and Impact on Human Health." *International Journal of Biological Innovations* 3(1): 38-47.

Preetha, B., and Viruthagiri, T. 2007. "Bioaccumulation of chromium (VI), copper (II) and nickel (II) ions by growing *Rhizopus arrhizus*." *Biochemical engineering journal* 34(2): 131-135.

Rebello, S., Anoopkumar, A. N., Aneesh, E. M., Sindhu, R., Binod, P., Kim, S. H., and Pandey, A. 2021. "Hazardous minerals mining: challenges and solutions." *Journal of Hazardous Materials* 402: 123474.

Reeves, R. D., Baker, A. J., Jaffré, T., Erskine, P. D., Echevarria, G., and van der Ent, A. 2018. "A global database for plants that hyperaccumulate metal and metalloid trace elements." *New Phytologist* 218(2): 407-411.

Reeves, R. D., van der Ent, A., Echevarria, G., Isnard, S., and Baker, A. J. 2021. "Global distribution and ecology of hyperaccumulator plants." In: *Agromining: Farming for metals, mineral resource reviews*. pp. 133-154. Springer, Cham.

Rice, K. M., Walker Jr, E. M., Wu, M., Gillette, C., and Blough, E. R. 2014. "Environmental mercury and its toxic effects." *Journal of Preventive Medicine and Public Health* 47(2): 74-83.

Şahan, T., Ceylan, H., Şahiner, N., and Aktaş, N. 2010. "Optimization of removal conditions of copper ions from aqueous solutions by *Trametes versicolor*." *Bioresource Technology* 101(12): 4520-4526.

Šalamún, P., Renčo, M., Kucanová, E., Brázová, T., Papajová, I., Miklisová, D., and Hanzelová, V. 2012. "Nematodes as bioindicators of soil degradation due to heavy metals." *Ecotoxicology* 21(8): 2319-2330.

Shrivastava, M., Khandelwal, A., and Srivastava, S. 2019. "Heavy metal hyperaccumulator plants: The resource to understand the extreme adaptations of plants towards heavy metals." In: *Plant-metal Interactions*. pp. 79-97. Springer, Cham.

Singh, A., Mehta, S. K., and Gaur, J. P. 2007. "Removal of heavy metals from aqueous solution by common freshwater filamentous algae." *World Journal of Microbiology and Biotechnology* 23(8): 1115-1120.

Stankovic, S., and Stankovic, A. R. 2013. "Bioindicators of toxic metals." Lichtfouse, E., Schwarzbauer, J., and Robert, D. (Eds.). In: *Green materials for energy, products and depollution.* pp. 151-228. Springer Netherlands.

Thacker, U., Parikh, R., Shouche, Y., and Madamwar, D. 2006. "Hexavalent chromium reduction by *Providencia* sp." *Process Biochemistry* 41(6): 1332-1337.

Tovar-Sánchez, E., Cervantes, L. T., Martínez, C., Rojas, E., Valverde, M., Ortiz-Hernández, M. L., and Mussali-Galante, P. 2012. "Comparison of two wild rodent species as sentinels of environmental contamination by mine tailings." *Environmental Science and Pollution Research* 19(5): 1677-1686.

Volke, S. T., Velasco, T. J. A., and De la Rosa, P. D. A. 2005. *Suelos contaminados por metales y metaloides: muestreo y alternativas para su remediación.* Secretaria de Medio Ambiente y Recursos Naturales e Instituto Nacional de Ecología. pp. 1-144. [*Soils contaminated by metals and metalloids: sampling and alternatives for their remediation.*]

Wiatrowski, H. A., Ward, P. M., and Barkay, T. 2006. "Novel reduction of mercury (II) by mercury-sensitive dissimilatory metal reducing bacteria." *Environmental Science & Technology* 40(21): 6690-6696.

Wu, H., Wu, Q., Wu, G., Gu, Q., and Wei, L. 2016. "Cd-resistant strains of *B. cereus* S5 with endurance capacity and their capacities for cadmium removal from cadmium-polluted water." *PloS one* 11(4): 1-25.

Yadav, K. K., Gupta, N., Kumar, V., and Singh, J. K. 2017. "Bioremediation of heavy metals from contaminated sites using potential species: a review." *Indian Journal of Environmental Protection* 37(1): 65-84.

Yang, S., Sun, X., Shen, Y., Chang, C., Guo, E., La, G., Zhao, Y., and Li, X. 2017. "Tolerance and removal mechanisms of heavy metals by fungus *Pleurotus ostreatus* HAAS." *Water, Air, & Soil Pollution* 228(4): 130-138.

Zukal, J., Pikula, J., and Bandouchova, H. 2015. "Bats as bioindicators of heavy metal pollution: history and prospect." *Mammalian Biology* 80(3): 220-227.

Biographical Sketch

Efraín Tovar Sánchez

Affiliation: Universidad Autónoma del Estado de Morelos

Education: PhD in Biological Sciences

Research and Professional Experience: Ecotoxicology

Professional Appointments: Senior Professor-Research

Publications from the Last 3 Years:

1. López-Caamal, A. and Tovar-Sánchez E. Accepted. Comparing the population history of Neotropical annual species: The role of climate change and hybridization between *Tithonia tubaeformis* and *T. rotundifolia* (Asteraceae). *Plant Biology.*
2. De la Cruz-Guarneros Natalia, Tovar-Sánchez Efraín and Mussali-Galante Patricia. 2021. Assessing effects of chronic heavy metal exposure through a multibiomarker approach: the case of *Liomys irroratus* (Rodentia: Heteromyidae). *Environmental Science and Pollution Research*, 1-15. https://doi.org/10.1007/s11356-021-14855-w.
3. Ortiz-Hernández Ma. Laura, Gama-Martínez Yitzel, Fernández-López Maikel, Castrejón-Godínez, María Luisa, Encarnación Sergio, Tovar-Sánchez Efraín, Salazar, E., Rodríguez, A., and Mussali-Galante Patricia. 2021. Transcriptomic analysis of *Burkholderia cenocepacia* CEIB S5-2 during methyl parathion degradation. *Environmental Science and Pollution Research*, 1-18. https://doi.org/10.1007/s11356-021-13647-6.
4. Valencia-Cuevas, Almendra Rodríguez-Domínguez, Patricia Mussali-Galante, Fernando Ramos-Quintana, Efraín Tovar-Sánchez. 2020. Influence of edaphic factors along an altitudinal gradient on a litter arthropod community in an Abies-Quercus forest in Mexico. *Acta Oecologica*. 108: doi: 10.1016/j.actao.2020.103609 (corresponding author)

5. Dalia A. Muro-González, Patricia Mussali-Galante, Leticia Valencia Cuevas, Karen Flores-Trujillo and Efraín Tovar-Sánchez. 2020. Morphological, physiological, and genotoxic effects of heavy metal bioaccumulation in *Prosopis laevigata* reveal its potential for phytoremediation. *Environmental Science and Pollution Research. 27*(32), 40187-40204. https://doi.org/10.1007/s11356-020-10026-5 (corresponding author).
6. Santoyo-Martínez, M., Mussali-Galante, P., Hernández-Plata, I., Valencia-Cuevas, L., Flores-Morales, A., Ortiz-Hernández, L., Tovar-Sánchez, E. 2020. Heavy metal bioaccumulation and morphological changes in *Vachellia campechiana* (Fabaceae) reveal its potential for phytoextraction of Cr, Cu, and Pb in mine tailings. *Environmental Science and Pollution Research*, 1-17. https://doi.org/10.1007/s11356-020-07730-7 (corresponding author)
7. Lobato-Vila, I., Cibrián-Tovar, D., Barrera-Ruíz, U. M., Equihua-Martínez, A., Estrada-Venegas, E. G., Tovar-Sánchez, E., Castillo-Mendoza, E., Buffington, M. L., and Pujade-Villar, J. (2020). Review of the Synergus Hartig species (Hymenoptera: Cynipidae: Synergini) associated with woolly galls on oaks from the New World, with the description of a new species from Mexico. *Proceedings of the Entomological Society of Washington,* 122(3), 678-715. DOI: 10.4289/0013-8797.122.3.678.
8. Flores-Trujillo, A. K. I., Mussali-Galante, P., Calero de Hoces, M., Blázquez-García, G., Saldarriaga-Noreña, H. A., Rodríguez-Solís, A., Tovar-Sánchez, E., Sánchez-Salinas E. and Ortiz-Hernández, L. 2020. Biosorption of heavy metals on *Opuntia fuliginosa* and *Agave angustifolia* fibers for their elimination from water. *International Journal of Environmental Science and Technology*. doi 10.1007/s13762-020-02832-8.
9. Hernández-Plata I., Rodríguez VM., Tovar-Sánchez E., Carrizalez L., Villalobos P., Mendoza-Trejo MS., Mussali-Galante P. 2020. Metal brain bioaccumulation and neurobehavioral effects on the wild rodent *Liomys irroratus* inhabiting mine tailing areas. *Environmental Science and*

Pollution Research. https://doi.org/10.1007/s11356-020-09451-3.

10. Salazar-Ramírez, G., Flores-Vallejo, R. D. C., Rivera-Leyva, J. C., Tovar-Sánchez, E., Sánchez-Reyes, A., Mena-Portales, J., Mussali-Galante, P. 2020. Characterization of Fungal Endophytes Isolated from the Metal Hyperaccumulator Plant *Vachellia farnesiana* Growing in Mine Tailings. *Microorganisms,* 8(2), 226. doi:10.3390/microorganisms8020226.
11. Rodríguez A., Castrejón-Godínez ML., Salazar-Bustamante E., Gama-Martínez Y., Sánchez-Salinas E., Mussali-Galante P., Tovar-Sánchez E., Ortiz-Hernández ML., 2020. Omics Approaches to Pesticide Biodegradation. *Current Microbiology* 77:545–563. doi: 10.10007/s00284-020-01916-5.
12. Ramírez-Rodríguez, R., Ocampo Bautista F., Rojas Flores B., Fores-Franco G., Tovar-Sánchez E. and Sánchez Popoca A. 2020. Flora arbórea no nativa, un potencial riesgo para la biodiversidad. In: La biodiversidad en Morelos: Estudio del Estado 2. Vol III. *Comisión Nacional para el Conocimiento de la Biodiversidad. Printed in Mexico.* Pp. 234-240. ISBN: 978-607-8570-42-3. [Non-native arboreal flora, a potential risk for biodiversity. In: *Biodiversity in Morelos: Study of the State 2. Vol III. National Commission for the Knowledge of Biodiversity. Printed in Mexico*]
13. Mussali-Galante P. and Tovar-Sánchez. E. 2020. Diversidad genética de especies centinela como bioindicador de la salud poblacional. In: La biodiversidad en Morelos: Estudio del Estado 2. Vol II. *Comisión Nacional para el Conocimiento de la Biodiversidad. Printed in Mexico.* Pp. 430-434. ISBN: 978-607-8570-41-6. [Genetic diversity of sentinel species as a bioindicator of population health. In: *Biodiversity in Morelos: Study of the State 2. Vol II. National Commission for the Knowledge of Biodiversity. Printed in Mexico.*]
14. Tovar-Sánchez, E. and Nava-García E. 2020. Aplicaciones del conocimiento sobre la diversidad genética. In: La biodiversidad en Morelos: Estudio del Estado 2. Vol II. *Comisión Nacional para el Conocimiento de la Biodiversidad.* Printed in Mexico. Pp. 425-429. ISBN: 978-607-8570-41-6. [Applications of

knowledge on genetic diversity. In: *Biodiversity in Morelos: Study of the State 2. Vol II. National Commission for the Knowledge of Biodiversity*]

15. Valencia-Cuevas, L., Castillo-Mendoza, E., Serrano-Muñoz- M., Tovar-Sánchez, E. 2020. Mexican oaks as foundation species: The case of *Quercus crassipes* and *Q. castanea*. In: Quercus: *Classification, Ecology and Uses*. Steffensen B. J. (ed.). pp. 211-242. Chapter 5. Nova Science Publishers Inc, USA. ISBN: 978-1-53618-026-8.
16. Castillo-Mendoza E., Mussali-Galante P., Valencia-Cuevas L., Flores Franco G., Castrejón Godínez L., Hernández-Plata I, Galván MA, Rodríguez Solís AJ and Tovar-Sánchez E. 2020. Half of Mexico's invasive plant species may be useful for heavy metal phytoremediation. In: Vinícius Londe (ed.). *Invasive species: Ecology, Impacts, and Potential Uses*. Chapter 8. Pp. 247-309. Nova Science Publishers Inc, USA. Pp. 247-309. ISBN: 978-1-53617-890-6.
17. Fernando Quintana, Efraín Tovar-Sánchez, Hugo Saldarriaga-Noreña, Héctor Sotelo-Nava, Juan Paulo Sánchez-Hernández, María-Luisa Castrejón-Godínez. 2019. A CBR-AHP Hybrid Method to Support the Decision-Making Process in the Selection of Environmental Management Actions. *Sustainability*. doi: 10.20944/preprints201909.0195.v1.
18. Ramos-Quintana, F., Héctor Sotelo-Nava, Hugo Saldarriaga-Noreña, Efraín Tovar-Sánchez. 2019. Assessing multifactorial human-nature interactions affecting the environmental quality through models of dynamic networks: a system thinking approach. *Sustainability*. 11:1957 doi:10.3390/su11071957.
19. Castrejón-Godínez, M.L., Sánchez-Salinas, E., Mussali-Galante, P., Salazar-Bustamante, E., Encarnación, S., Rodríguez, A., Tovar-Sánchez. E. 2019. Transcriptional analysis reveals the metabolic state of *Burkholderia zhejiangensis* CEIB S4-3 during methyl parathion degradation. *International Biodeterioration & Biodegradation*. PeerJ. doi: 10.7717/peerj.6822.

20. López-Caamal A., Ferrufino-Acosta L.F., Díaz-Maradiaga R.F., Rodríguez-Delcid, D., Mussali-Galante P., Tovar-Sánchez E. 2019. Species distribution modelling and cpSSR reveal population history of the Neotropical annual herb *Tithonia rotundifolia* (Asteraceae). *Plant Biology.* doi: 10.1111/plb.12925 (corresponding author).
21. Tovar-Sánchez, E., Suarez-Rodríguez, R., Ramírez-Trujillo, A., Valencia-Cuevas, L., Hernández-Plata, I., and Mussali-Galante, P. 2019. The use of biosensors for biomonitoring environmental metal pollution. In: Rinken T. and Kivira K (eds), *Environmental Biosensors,* IntechOpen, London, UK. ISBN: 978-953-307-486-3. doi: http://dx.doi.org/10.5772/intechopen.84309.

In: Environmental Management
Editor: Miguel Fischer
ISBN: 978-1-68507-019-9

Chapter 5

GUIDELINES FOR SOLID WASTE MANAGEMENT: OVERCOMING CHALLENGES AND EXPLOIT OPPORTUNITIES

R. G. Anuardo*, M. Espuny†, MD, V. de M. Oliveira‡ and O. J. de Oliveira§, PhD
Production Department, São Paulo State University,
Guaratinguetá, São Paulo, Brazil

ABSTRACT

Approximately one million people die each year in the world because of chemical contamination caused by solid waste disposal. From 1991 to 2000, the amount of solid waste produced in the world was 0.68 billion tons, from 2010 to 2018 it increased to 1.3 billion tons, and it is estimated that by 2025 it will be 2.2 billion tons. Solid waste includes food waste, paper, plastics, glass, textiles, metals, wood, leather, and others; and is produced in human activities, both in households and organizations, and

* Corresponding Author's E-mail: r.anuardo@hotmail.com.
† Corresponding Author's E-mail: maximilian.espuny@unesp.br.
‡ Corresponding Author's E-mail: moraes.oliveira@unesp.br.
§ Corresponding Author's E-mail: otaviodeoliveira@uol.com.br.

requires handling, storage, collection, and disposal. The main challenges identified in the literature on solid waste management are difficulty in measuring and controlling environmental and social impacts, lack of legislation applied to waste and of specialized workers, structural and budgetary deficiency, and lack of management-oriented strategies and statistical tools. Within this scenario, the objective of this chapter is to propose recommendations for developing solid waste management based on the identification of scientific opportunities and challenges in the literature. Based on what was identified, benchmarking was performed, and recommendations were proposed to overcome the challenges of solid waste management. The main scientific contribution of this chapter is the deepening and expansion of the literature on solid waste management based on the solutions to the challenges, which will help and stimulate future research on the topic. As for the applied contribution, this work presents suggestions of actions for public managers and entrepreneurs of the solid waste sector with the perspective of process automation, waste destination, sustainability of the planet, and the reduction of waste production.

Keywords: solid waste management, waste, recommendations, challenges

INTRODUCTION

Solid waste management (SWM) has become increasingly complex as the population increases in urban regions and consumption habits promote heterogeneous waste disposal to landfills (Johari et al. 2014). The lack of organization of correct disposal brings people in direct contact with waste, contaminating and killing about 1.5 million people annually (ONU 2015). To reduce the impact of lethality caused by inadequate waste management, SWM planning and implementation are necessary, directly intervening in the reduction, proper treatment, and energy use of waste (Tan et al. 2014). Governments should establish guidelines to guide SWM, so that people and organizations fulfill their obligations related to reducing the generation and proper disposal of waste (Ferri, Diniz Chaves, and Ribeiro 2015; Herva, Neto, and Roca 2014; Li and Zhao 2015). To fulfill this role, the government must rely on the support of non-governmental organizations; the technology industry capable of developing equipment that encourages waste treatment efficiency; educational institutions from basic to higher levels to raise awareness of research to help solve SWM challenges; and other stakeholders (Cucchiella, D'Adamo, and Gastaldi 2014; Liu et al. 2014).

Solid waste can be classified as all discarded items regardless of their intrinsic value, upon the decision of its owner to throw it away (Johari et al. 2014). Solid waste encompasses "food waste, paper, cardboard, plastics, PET, glass, textiles, metals, wood and leather, diapers, ash and among others" (Tozlu, Özahi, and Abuşołlu 2016). Solid waste is produced in human activities, both in homes and in organizations, and its handling, storage, collection, and disposal are essential in order not to bring inconvenience to people. This process in all its completeness, if not adequate, can leave the environment and the population vulnerable to toxicities (Gupta, Yadav, and Kumar 2015).

According to Complementary Law 12,350/2010, waste can be classified as hazardous or non-hazardous (Brasil 2010). The first category is commonly referred to as "waste," collected from systematization made by communities. Hazardous materials include infectious materials, which are usually incinerated; and toxic waste, such as nuclear waste that cannot be exposed to the public (Beliën, De Boeck, and Van Ackere 2012). Recyclable waste includes paper, plastic, glass, metals, among others.

To reduce the amount of solid waste in nature it is important to have a multidisciplinary effort, aiming at the protection of the environment and the conservation of natural resources through an integrated waste management system with a focus on sustainable urban development (Cucchiella, D'Adamo, and Gastaldi 2014; H.-C. Liu et al. 2014).

The literature shows a concentration of studies focused on management tools, laws and regulations, and solid waste treatment (Ferri, Diniz Chaves, and Ribeiro 2015; Herva, Neto, and Roca 2014; Li and Zhao 2015). Management tools contribute to improving the results of solid waste management by increasing the efficiency of processes aimed at meeting the established goals (Herva, Neto, and Roca 2014).

Decreasing reliance on landfills and dumpsites is critical to adequate SW management. Failure to do so can culminate in the degradation of precious resources, as well as raise the cost of land and subject the environment and the human species to long-term weathering (Tan et al. 2014). According to Tan et al. 2015, "landfilling is a cheap technique for dealing with waste in large quantities," although there is opposition from society and a lack of availability of spaces to carry out this practice.

To no longer be dependent on dumps and landfills, it is necessary that waste management has integrated, in the sequence, subjecting the waste to a fractioning and then forwarding it to an effective treatment process. Among the most traditional techniques is the selection and separation, however, what makes the method successful is the active participation of the members who are waste generators (Mensah et al. 2015).

With better sorted and separated waste, the potential for revenue generation increases considerably and can transform an average environment with a rate between 5-10% of collected waste into higher percentages, assuming that the recyclable ejected elements are in the order of 70-80%. The current reality is a 90-95% rate of solid waste disposed of in landfills (Johari et al. 2014). Considering the estimated increase to 2.2 billion tons (ONU 2015) if this high rate remains in landfills, environmental degradation is expected to be even more pronounced.

Despite the improvement of the most diverse areas focused on solid waste management, from the field of technology to the legal field, the public acceptance to house treatment facilities is tiny (Mir et al. 2016). The process of replacing landfills and dumps with other techniques requires a broad discussion from the authorities. The population, in general, opposes the implementations of landfills in their perimeter, providing a social phenomenon known as "not in my backyard" (Liu et al. 2014).

Despite SWM being indispensable for the population's quality of life and an essential tool for environmental preservation, governments at various levels and in many regions of the planet have identified several difficulties in implementing guidelines that are enforceable and technical procedures that are enforceable. These problems demand constant challenges and propositions of solutions to overcome them. Given the above, "what are the main opportunities and challenges for the development of solid waste management?" is the question of this book chapter. To answer it, the aim of this work is to propose recommendations for the development of SWM based on the identification of scientific opportunities and challenges. For the development of the recommendations, content analysis was used from articles indexed in the Scopus database.

In addition to this introduction, this book chapter has the method section; results, and discussion containing the grouping of scientific opportunities and challenges identified in the literature, proposed recommendations; conclusions, and references.

Research Method

This topic describes the research method used to develop this chapter, following the methodological flow presented in Figure 1.

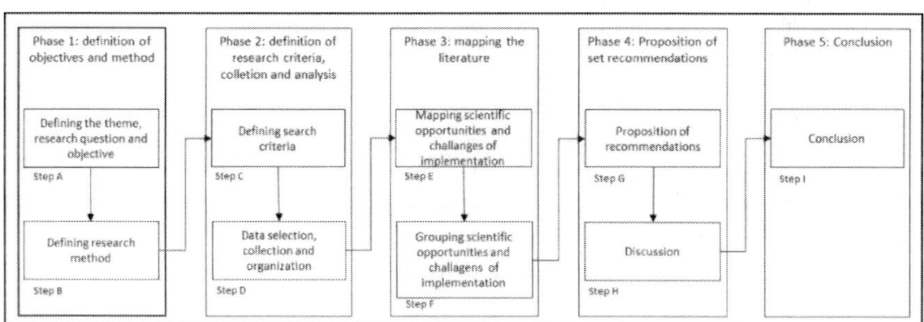

Figure 1. Methodological flow.

The procedures of the methodological flow are described in Table 1. Following the five development stages and their respective steps.

Table 1. Description of the methodological flow

Phases	Steps	Description
Phase 1	A: Defining the theme, research question, and objective	The research theme was delimited. The research question and the objective that would guide the execution of the chapter was conducted
	B: Research method	The method used to obtain the research results was defined, executing the objectives and answer the research question
Phase 2	C: Defining search criteria	The search criteria were defined in the Scopus database. This database was preferred because it is one of the most complete databases for subjects related to management (Santos et al. 2020). Only articles in English were chosen because they are the most widely used in the scientific environment (Nunhes and Oliveira 2018). The search query consulted was "solid waste management" only in the title of the articles. The period covered was from 2016 to 2021.
	D: Data selection, collection, and organization	The search return, according to the query mentioned in the previous step, identified 1,303 articles. From this total, the 30 most cited articles were selected to be used in the next steps. These articles served as a basis for the clustering of scientific opportunities and the main challenges. The information was grouped according to Tables 2 and 3.
Phase 3	E: Mapping scientific opportunities and challenges of implementation	The mapping of scientific opportunities and challenges was performed using content analysis, which is an appropriate method for identifying patterns in subjective analyses. This method enables the understanding of phenomena and the construction of knowledge through the information used (Elo, Forman, and Damschroder 2007; Hsieh and Shannon 2005).
	F: Grouping scientific opportunities and challenges	The opportunities and challenges of SMW were identified by performing content analysis, according to the similarities of their attributes.
Phase 4	G: Proposition of recommendations	The propositions of the recommendations were developed based on the experience of the authors, added to benchmarking of the opportunities and challenges identified in the scientific literature (Figure 2).
	H: Discussion	The recommendations were discussed considering the authors' experience and the theoretical conceptions of SWM, to allow for the feasibility of their implementation.
Phase 5	I: Conclusion	The paper was concluded, reporting the fulfillment of the objectives, highlighting its contributions, study limitations, and suggestions for future studies in SWM.

RESULTS

This topic presents the scientific opportunities of SWM and the challenges for its development. The scientific opportunities and challenges in solid waste management were identified in the 30 most cited articles on the topic between 2016 and 2020.

Clustering of Scientific Opportunities

The scientific opportunities identified in the literature were grouped according to their common characteristics, that is, considering their nature, resulting in the clusters presented in Table 2.

The first cluster addresses the need for solutions to implement and develop SWM practices. The SWM is a global trend due to the growing demand for waste produced annually. However, this field is still in its infancy, therefore, in constant improvement. Because of this, it is necessary to develop and improve solutions that support the development, promotion, and improvement of its management, seeking cheaper, efficient, effective, and easy to apply solutions.

Table 2. Clustering of scientific opportunities identified in the literature

Clusters	Authors
Propose solutions for the implementation and development of SWM practices	Vučijak, Kurtagić, and Silajdžić (2016), Ripa et al.(2017), Liu, Xing, and Liu (2017), Nabavi-Pelesaraei et al. (2017), Vitorino de Souza Melaré et al. (2017), Coban, Ertis, and Cavdaroglu (2018), Cremiato et al. (2018), Khandelwal et al. (2019)
Identify and systematize government practices that foment SWM	Joshi and Ahmed (2016), Ma and Hipel (2016), Fei et al. (2016), Moh and Abd Manaf (2017), Aparcana (2017), Kamaruddin et al. (2017), Rodić and Wilson (2017), Chien Bong et al. (2017)
Evaluate possible improvements and sustainable alternatives for SWM processes	Mir et al. (2016), Tozlu, Özahi, and Abuşoğlu (2016), Turner, Williams, and Kemp (2016), Sukholthaman and Sharp (2016), Sadef et al. (2016), Asefi and Lim (2017), Havukainen et al. (2017), Nabavi-Pelesaraei, Bayat, Hosseinzadeh-Bandbafha, Afrasyabi, and Berrada (2017), Mian et al. (2017), Fernández-González et al. (2017), Jara-Samaniego et al. (2017), Das et al. (2019)

The second cluster of research gaps addresses the need to develop legislation, policies, and government programs that foster the development of SW practices and reduce informal labor. Thus seeking to reduce environmental impacts, damage to the health of informal workers, increase the adherence of society, and the success of SWM.

The third cluster seeks the development of improvements and alternatives for SWM processes. Several solutions currently used are unsustainable, such as SW incineration, or need process improvements, such as biogas production, and the tools and models for calculating waste control are difficult to apply because it is a dynamic element with a variation of its composition.

Clustering of Challenges of SWM

Solid waste management challenges were identified in the 30 most cited articles on the topic between 2016 and 2020. They were then grouped according to their similarities (Table 3).

Table 3. Clusters of solid waste management challenges

Clusters	Authors
Difficulty in measuring and controlling environmental and social impacts	Fei et al. (2016), Sadef et al. (2016), Malinauskaite et al. (2017), Nabavi-Pelesaraei, Bayat, Hosseinzadeh-Bandbafha, Afrasyabi, and Berrada (2017), Fernández-González et al. (2017)
Lack of legislation applied to waste and of specialized workers	Turner, Williams, and Kemp (2016), Joshi and Ahmed (2016), Ma and Hipel (2016), Mian et al. (2017), Aparcana (2017), Rodić and Wilson (2017), Das et al. (2019)
Structural and budgetary deficiency	Tozlu, Özahi, and Abuşoğlu (2016), Sukholthaman and Sharp (2016), Ripa et al.(2017), Vitorino de Souza Melaré et al. (2017), Havukainen et al. (2017), Jara-Samaniego et al. (2017), Chien Bong et al. (2017), Nabavi-Pelesaraei et al. (2017), Cremiato et al. (2018)
Lack of management-oriented strategies and statistical tools	Mir et al. (2016), Vučijak, Kurtagić, and Silajdžić (2016), Rajaeifar et al. (2017), Moh and Abd Manaf (2017), Kamaruddin et al. (2017), Liu, Xing, and Liu (2017), Asefi and Lim (2017), Coban, Ertis, and Cavdaroglu (2018), Khandelwal et al. (2019)

The cluster "Difficulty in measuring and controlling environmental and social impacts" contemplates the lack of regulation and control of the solid waste management structure and its informal workers, leading to negative environmental and social impacts. According to Nabavi-Pelesaraei, Bayat, Hosseinzadeh-Bandbafha, Afrasyabi, and Chau (2017), the solid waste landfill system is a serious threat to the environment due to the lack of complete control of the systems and the impact of these threats.

The challenges of the cluster "Lack of legislation applied to waste and of specialized workers" highlight the low effectiveness in solid waste management due to informality and insufficient enforcement of laws directed to its disposal. According to Rodić and Wilson (2017), the implementation of these laws is mostly inadequate due to limited technical and financial capacity within municipal administrations.

The "Structural and budgetary deficiency" cluster represents inadequate solid waste management structure and the difficulty of using more sustainable treatments and services due to the financial resources available. According to Chien Bong et al. (2017), one way to manage solid waste and obtain renewable energy is through biogas production. However, the financial cost, unsafe feedstock, infrastructure requirements, and technical maturity hinder the expansion of its production.

The cluster "Lack of management-oriented strategies and statistical tools" indicates that the articles that make it up to have identified inadequate solid waste management. This management presents obstacles to deciding and realizing the most appropriate method for waste collection and treatment. According to Mir et al. (2016), several methods can be applied to manage solid waste, such as thermal treatment and landfills. However, it is necessary to select the most suitable SW treatment method that provides the greatest economic and environmental benefits.

PROPOSED RECOMMENDATION AND DISCUSSION

From the content analysis performed on the 30 main articles identified in the literature about SWM, we formulated 11 recommendations for its improvement, as shown in Figure 2: (I) Create government programs that stimulate the participation of private companies in solid waste management; (II) Formalize and equip recycling cooperatives in order to generate jobs and reduce the amount of recyclable waste in landfills; (III) Instituting technical, undergraduate and graduate courses that address the core issues of waste management; (IV) Prepare waste management teams capable of planning and executing waste management programs; (V) Create tools for the quantification and control of solid waste; (VI) Develop training programs for professionals working in SWM; (VII) Create disposal sites at strategic points to facilitate the population's access; (VIII) Make feasible the adoption of more sustainable methods such as recycling, composting and incineration as an alternative to the use of landfills; (IX) Implement more efficient methods of enforcement of the laws related to solid waste; (X) Create laws that encourage the adequate disposal of solid waste by the population; and (XI) Identify alternative methods of treating solid waste because of the heterogeneity of its composition.

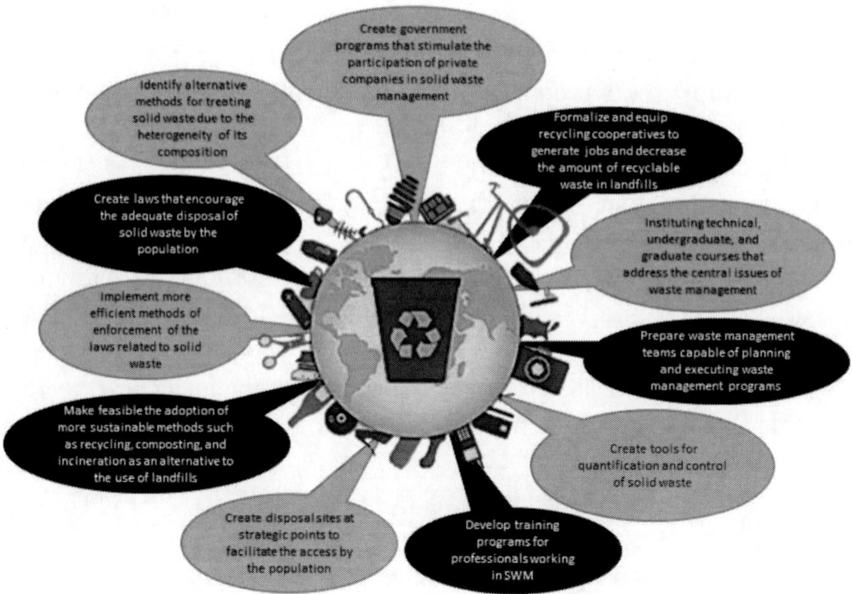

Figure 2. Proposed recommendations for SWM development.

Regarding the creation of government programs (I), the marginality of waste management has been a serious problem for its administration, due to the costs for its treatment and few incentives from governments to promote SWM. Therefore, the creation of laws and programs that encourage this participation, either through financial or technological resources, helps in the management of solid waste. Additionally, benefits can be granted to companies that practice and promote SWM in their facilities and to their employees.

As for the formalization and provision of equipment to cooperatives (II), it is important to highlight that due to the cost of waste control and treatment, formal systems generally have small amounts of recycling. This leads to the dumping of recyclable waste in landfills, a large presence of informal collectors, and the small participation of formal recycling cooperatives. These cooperatives are characterized by small-scale, low-tech, low-paid, unregistered, unregulated work, and operated with low capital investment. Therefore, it is necessary to gradually transfer the participants of these cooperatives into the formal system, standardizing their operations and equipping them, which leads to job creation and improvement of the workers who came from the informal sector (Fei et al. 2016).

Another critical point for developing SWM is the lack of trained professionals (III). It is the government's responsibility to invest in the training of trained professionals to apply SWM and promote the development of new solutions. For this, there are proposed actions to train entrepreneurs to develop SWM in their enterprises by minimizing SW and investing in research to develop solutions for implementation, analysis, evaluation, quantification, and treatment of SW.

As for the preparation of management teams in SWM (IV), it is important to emphasize that federal-level laws tend to present guidelines for waste management, but, it is the municipalities that execute them. In developed municipalities with high budgets, it is possible to appoint managers with environmental and SWM expertise, but in a more remote municipalities or countries with few resources, the specialization of these officials becomes more difficult. To remedy this difficulty, the authorities above the municipalities should prepare SWM-oriented staff members in their centralized units. Among the skills that the staff members should have are the identification of waste generation sources, waste transportation and collection, waste treatment, and final disposal.

Among the proposed recommendations, we cannot fail to mention quantitative tools to assist in the control of SWM (V). Due to the heterogeneous characteristics of solid waste, the government must seek solutions that provide precise characteristics of the waste, indicating the changes in composition that it undergoes in a given period. Tools that can fulfill this role well are SWOT analyses, which combine qualitative and quantitative tools to support decisions by policymakers (Zorpas 2020). Other methods that are highly encouraged are cost-benefit analysis (CBA) and life cycle assessment (LCA), which are commonly used for cost-benefit comparison in the steps employed in SWM (Pan et al. 2015).

Another important point to be recommended is to improve the training of those directly involved in the waste activity (VI). SWM professionals are often socially marginalized and have little training in their work. By developing training programs for these professionals, there is a greater specialization of their work, contributing to an improvement in the quality of waste management, an improvement in the lives of workers, and a more sustainable planet. Besides the waste cooperatives, several branches that can still be explored in this market. Some examples that can be cited are a waste brokerage, in which professionals can earn a percentage of the transaction values between buyer and seller, and urban mining, in which professionals can extract gold, silver, and other valuable metals from technology equipment, such as computers and cell phones.

The physical structures for waste disposal should be part of a plan to improve SWM (VII). In this context, the difficulty of access to disposal sites ends up stimulating the population to dispose of SW incorrectly, promoting the informality of SWM through the collectors and increasing the chances of contamination. Thus, the creation of strategic disposal points would facilitate access, along with the stimulation of awareness, which will contribute to the reduction of the incorrect and informal management of SW.

The incorrect management mentioned in the previous paragraph, creates conditions for landfills to be prioritized, even though they are the worst alternative for SWM (VIII). Landfills are the worst alternative that SWM must resort to. In the last two decades, public managers have sought to implement laws and incentive campaigns so that this trend can be reversed. To achieve this goal, there must be an incentive from the public sector to develop technologies capable of mitigating the simple displacement of waste to the disposal area. This initiative can create conditions for the waste to be recycled. In regions where coal is used, incineration is still encouraged as an option for energy generation.

For SWM to impact society, including those recommended in this section, everyone must play their role properly, including citizens and organizations. If the laws imposed by the government are not properly enforced, there is a risk that citizens will not be motivated to deposit the allowed amount of waste, and organizations may manipulate the weight of the waste going to landfills, circumventing what is foreseen in the laws (IX). If the waste inspection is well employed, there will be an environmental return, or, in the worst-case scenario, there will be a financial reimbursement due to the fines.

Besides enforcement itself, governments should encourage the population to take initiatives from the "bottom-up," as well. This could be done through the creation of laws to encourage the disposal of solid waste (X) in more appropriate locations by citizens is necessary because with this there is less SW in landfills and greater urban cleanliness. This incentive can be through the teaching of correct disposal in schools, advertising actions such as videos and posters, and the availability of volunteers who assist in this disposal at collection points. It is also possible to provide financial incentives for those who correctly dispose of SW, such as credits that can be exchanged for food products and theater tickets (Zhou et al. 2019).

Finally, the last recommendation is about the heterogeneity of solid waste hinders its treatment and use as a raw material for recycling, requiring a greater fractioning of this waste (XI). With this appropriate fractioning in the collection, an analysis must be conducted of which is the most appropriate and sustainable treatment for this type of SW, avoiding its disposal in landfills. Among these treatments are anaerobic digestion, biogas, composting, and recycling.

Conclusion

The rural exodus has caused an increasing concentration of people in urban areas, which has required the mayors of all major cities in the world to think of alternatives to improve municipal SWM. What has been noticed is that very few municipalities have been able to overcome these challenges, which ends up demanding a lot of effort from companies, citizens, companies specialized in SWM, and the public power itself. Considering these difficulties, this chapter proposed recommendations that could contribute to public and private managers for a significant change in this scenario. Throughout the manuscript, the opportunities, and challenges of SWM were systematized and then served as a basis for the formulation of recommendations.

The deepening and expansion of the literature were the main academic contributions provided by this chapter, through the recommendations provided to overcome the challenges of SWM. As for the applied contribution, one can rule out the recommendations that can be used by governments and businessmen to improve SWM, benefiting the environment, society, and the economy.

Regarding the main limitations of this study, one can mention that it was the fact that it was conducted only based on theory, failing to portray empirical practices that could be incorporated into the theory of SWM. Therefore, such studies that consider the waste management scenario mainly in municipalities, can validate the proposed propositions and even expand the quality of the propositions.

Acknowledgments

This study was funded in part by the Coordenação de Aperfeiçoamento de Pessoal de Nível Superior – Brazil (CAPES) – Financial Code 001, CNPq – Conselho Nacional de Desenvolvimento Científico Tecnológico – (312894/2017-1 and 312538/2020-0) and the Fundação de Amparo à Pesquisa do Estado de São Paulo (FAPESP) [Grant number 2019/23998-3] for financial support.

References

Aparcana, Sandra. 2017. “Approaches to Formalization of the Informal Waste Sector into Municipal Solid Waste Management Systems in Low- and Middle-Income Countries: Review of Barriers and Success Factors.” *Waste Management* 61: 593–607. https://doi.org/10.1016/j.wasman.2016.12.028.

Asefi, Hossein, and Samsung Lim. 2017. “A Novel Multi-Dimensional Modeling Approach to Integrated Municipal Solid Waste Management.” *Journal of Cleaner Production* 166: 1131–43. https://doi.org/10.1016/j.jclepro.2017.08.061.

Beliën, Jeroen, Liesje De Boeck, and Jonas Van Ackere. 2012. “Municipal Solid Waste Collection and Management Problems: A Literature Review.” *Transportation Science* 48 (1): 78–102. https://doi.org/10.1287/trsc.1120.0448.

Brasil. 2010. *Política Nacional de Resíduos Sólidos*. Brasil. http://www2.camara.leg.br/legin/fed/lei/2010/lei-12305-2-agosto-2010-%0A607598-norma-pl.html. [*National Solid Waste Policy*]

Chien Bong, Cassendra Phun, Wai Shin Ho, Haslenda Hashim, Jeng Shiun Lim, Chin Siong Ho, William Soo Peng Tan, and Chew Tin Lee. 2017. “Review on the Renewable Energy and Solid Waste Management Policies towards Biogas Development in Malaysia.” *Renewable and Sustainable Energy Reviews* 70 (December): 988–98. https://doi.org/10.1016/j.rser.2016.12.004.

Coban, Asli, Irem Firtina Ertis, and Nur Ayvaz Cavdaroglu. 2018. “Municipal Solid Waste Management via Multi-Criteria Decision Making Methods: A Case Study in Istanbul, Turkey.” *Journal of Cleaner Production* 180: 159–67. https://doi.org/10.1016/j.jclepro.2018.01.130.

Cremiato, Raffaele, Maria Laura Mastellone, Carla Tagliaferri, Lucio Zaccariello, and Paola Lettieri. 2018. “Environmental Impact of Municipal Solid Waste Management Using Life Cycle Assessment: The Effect of Anaerobic Digestion, Materials Recovery and Secondary Fuels Production.” *Renewable Energy* 124: 180–88. https://doi.org/10.1016/j.renene.2017.06.033.

Cucchiella, Federica, Idiano D’Adamo, and Massimo Gastaldi. 2014. “Strategic Municipal Solid Waste Management: A Quantitative

Model for Italian Regions." *Energy Conversion and Management* 77: 709–20. https://doi.org/10.1016/j.enconman.2013.10.024.

Das, Subhasish, S. H. Lee, Pawan Kumar, Ki Hyun Kim, Sang Soo Lee, and Satya Sundar Bhattacharya. 2019. "Solid Waste Management: Scope and the Challenge of Sustainability." *Journal of Cleaner Production* 228: 658–78. https://doi.org/10.1016/j.jclepro.2019.04.323.

Elo, Satu, Jane Forman, and Laura Damschroder. 2007. "Qualitative Content Analysis." *Advances in Bioethics* 11: 39–62. https://doi.org/10.1016/S1479-3709(07)11003-7.

Fei, Fan, Lili Qu, Zongguo Wen, Yanyan Xue, and Huanan Zhang. 2016. "How to Integrate the Informal Recycling System into Municipal Solid Waste Management in Developing Countries: Based on a China's Case in Suzhou Urban Area." *Resources, Conservation and Recycling* 110: 74–86. https://doi.org/10.1016/j.resconrec.2016.03.019.

Fernández-González, J. M., A. L. Grindlay, F. Serrano-Bernardo, M. I. Rodríguez-Rojas, and M. Zamorano. 2017. "Economic and Environmental Review of Waste-to-Energy Systems for Municipal Solid Waste Management in Medium and Small Municipalities." *Waste Management* 67: 360–74. https://doi.org/10.1016/j.wasman.2017.05.003.

Ferri, Giovane Lopes, Gisele de Lorena Diniz Chaves, and Glaydston Mattos Ribeiro. 2015. "Reverse Logistics Network for Municipal Solid Waste Management: The Inclusion of Waste Pickers as a Brazilian Legal Requirement." *Waste Management* 40: 173–91. https://doi.org/10.1016/j.wasman.2015.02.036.

Gupta, Neha, Krishna Kumar Yadav, and Vinit Kumar. 2015. "A Review on Current Status of Municipal Solid Waste Management in India." *Journal of Environmental Sciences (China)* 37: 206–17. https://doi.org/10.1016/j.jes.2015.01.034.

Havukainen, Jouni, Mingxiu Zhan, Jun Dong, Miia Liikanen, Ivan Deviatkin, Xiaodong Li, and Mika Horttanainen. 2017. "Environmental Impact Assessment of Municipal Solid Waste Management Incorporating Mechanical Treatment of Waste and Incineration in Hangzhou, China." *Journal of Cleaner Production* 141: 453–61. https://doi.org/10.1016/j.jclepro.2016.09.146.

Herva, Marta, Belmira Neto, and Enrique Roca. 2014. "Environmental Assessment of the Integrated Municipal Solid Waste Management System in Porto (Portugal)." *Journal of Cleaner Production* 70: 183–93. https://doi.org/10.1016/j.jclepro.2014.02.007.

Hsieh, Hsiu Fang, and Sarah E. Shannon. 2005. "Three Approaches to Qualitative Content Analysis." *Qualitative Health Research* 15 (9): 1277–88. https://doi.org/10.1177/1049732305276687.

Jara-Samaniego, J., M. D. Pérez-Murcia, M. A. Bustamante, A. Pérez-Espinosa, C. Paredes, M. López, D. B. López-Lluch, I. Gavilanes-Terán, and R. Moral. 2017. "Composting as Sustainable Strategy for Municipal Solid Waste Management in the Chimborazo Region, Ecuador: Suitability of the Obtained Composts for Seedling Production." *Journal of Cleaner Production* 141: 1349–58. https://doi.org/10.1016/j.jclepro.2016.09.178.

Johari, Anwar, Habib Alkali, Haslenda Hashim, Saeed I. Ahmed, and Ramli Mat. 2014. "Municipal Solid Waste Management and Potential Revenue from Recycling in Malaysia." *Modern Applied Science* 8 (4): 37–49. https://doi.org/10.5539/mas.v8n4p37.

Joshi, Rajkumar, and Sirajuddin Ahmed. 2016. "Status and Challenges of Municipal Solid Waste Management in India: A Review." *Cogent Environmental Science* 2 (1): 1–18. https://doi.org/10.1080/23311843.2016.1139434.

Kamaruddin, Mohamad Anuar, Mohd Suffian Yusoff, Lo Ming Rui, Awatif Md Isa, Mohd Hafiz Zawawi, and Rasyidah Alrozi. 2017. "An Overview of Municipal Solid Waste Management and Landfill Leachate Treatment: Malaysia and Asian Perspectives." *Environmental Science and Pollution Research* 24 (35): 26988–20. https://doi.org/10.1007/s11356-017-0303-9.

Khandelwal, Harshit, Hiya Dhar, Arun Kumar Thalla, and Sunil Kumar. 2019. "Application of Life Cycle Assessment in Municipal Solid Waste Management: A Worldwide Critical Review." *Journal of Cleaner Production* 209: 630–54. https://doi.org/10.1016/j.jclepro.2018.10.233.

Li, Wei, and Yang Zhao. 2015. "Bibliometric Analysis of Global Environmental Assessment Research in a 20-Year Period." *Environmental Impact Assessment Review* 50: 158–66. https://doi.org/10.1016/j.eiar.2014.09.012.

Liu, Hu-Chen, Jian-Xin You, Yi-Zeng Chen, and Xiao-Jun Fan. 2014. “Site Selection in Municipal Solid Waste Management with Extended VIKOR Method under Fuzzy Environment.” *Environmental Earth Sciences* 72 (10): 4179–89. https://doi.org/10.1007/s12665-014-3314-6.

Liu, Hu Chen, Jian Xin You, Xiao Jun Fan, and Yi Zeng Chen. 2014. “Site Selection in Waste Management by the VIKOR Method Using Linguistic Assessment.” *Applied Soft Computing Journal* 21: 453–61. https://doi.org/10.1016/j.asoc.2014.04.004.

Liu, Yili, Peixuan Xing, and Jianguo Liu. 2017. “Environmental Performance Evaluation of Different Municipal Solid Waste Management Scenarios in China.” *Resources, Conservation and Recycling* 125 (March): 98–106. https://doi.org/10.1016/j.resconrec.2017.06.005.

Ma, Jing, and Keith W. Hipel. 2016. “Exploring Social Dimensions of Municipal Solid Waste Management around the Globe – A Systematic Literature Review.” *Waste Management* 56: 3–12. https://doi.org/10.1016/j.wasman.2016.06.041.

Malinauskaite, J., H. Jouhara, D. Czajczyńska, P. Stanchev, E. Katsou, P. Rostkowski, R. J. Thorne, et al. 2017. “Municipal Solid Waste Management and Waste-to-Energy in the Context of a Circular Economy and Energy Recycling in Europe.” *Energy* 141: 2013–44. https://doi.org/10.1016/j.energy.2017.11.128.

Mensah, Moses Y., Bernard Fei-Baffoe, Kwasi Obiri-Danso, Kodwo Miezah, and Zsófia Kádár. 2015. “Municipal Solid Waste Characterization and Quantification as a Measure towards Effective Waste Management in Ghana.” *Waste Management* 46: 15–27. https://doi.org/10.1016/j.wasman.2015.09.009.

Mian, Md Manik, Xiaolan Zeng, Allama al Naim Bin Nasry, and Sulala M. Z. F. Al-Hamadani. 2017. “Municipal Solid Waste Management in China: A Comparative Analysis.” *Journal of Material Cycles and Waste Management* 19 (3): 1127–35. https://doi.org/10.1007/s10163-016-0509-9.

Mir, M Aghajani, P Taherei Ghazvinei, N M N Sulaiman, N E A Basri, S Saheri, N Z Mahmood, A Jahan, R A Begum, and N Aghamohammadi. 2016. “Application of TOPSIS and VIKOR Improved Versions in a Multi Criteria Decision Analysis to Develop an Optimized Municipal Solid Waste Management Model." *Journal of Environmental Management* 166: 109–15. https://doi.org/10.1016/j.jenvman.2015.09.028.

Moh, Yiing Chiee, and Latifah Abd Manaf. 2017. "Solid Waste Management Transformation and Future Challenges of Source Separation and Recycling Practice in Malaysia." *Resources, Conservation and Recycling* 116 (2017): 1–14. https://doi.org/10.1016/j.resconrec.2016.09.012.

Nabavi-Pelesaraei, Ashkan, Reza Bayat, Homa Hosseinzadeh-Bandbafha, Hadi Afrasyabi, and Asmae Berrada. 2017. "Prognostication of Energy Use and Environmental Impacts for Recycle System of Municipal Solid Waste Management." *Journal of Cleaner Production* 154: 602–13. https://doi.org/10.1016/j.jclepro.2017.04.033.

Nabavi-Pelesaraei, Ashkan, Reza Bayat, Homa Hosseinzadeh-Bandbafha, Hadi Afrasyabi, and Kwok wing Chau. 2017. "Modeling of Energy Consumption and Environmental Life Cycle Assessment for Incineration and Landfill Systems of Municipal Solid Waste Management - A Case Study in Tehran Metropolis of Iran." *Journal of Cleaner Production* 148: 427–40. https://doi.org/10.1016/j.jclepro.2017.01.172.

Nunhes, Thaís Vieira, and Otávio José Oliveira. 2018. "Analysis of Integrated Management Systems Research: Identifying Core Themes and Trends for Future Studies." *Total Quality Management & Business Excellence* 0 (0): 1–23. https://doi.org/10.1080/14783363.2018.1471981.

ONU. 2015. *ONU Prevê Que Mundo Terá 50 Milhões de Toneladas de Lixo Eletrônico Em 2017*. 2015. https://nacoesunidas.org/onu-preve-que-mundo-tera-50-milhoes-de-toneladas-de-lixo-eletronico-em-2017/. [*UN Predicts The World Will Have 50 Million Tonnes Of E-Waste By 2017*]

Pan, Shu Yuan, Michael Alex Du, I. Te Huang, I. Hung Liu, E. E. Chang, and Pen Chi Chiang. 2015. "Strategies on Implementation of Waste-to-Energy (WTE) Supply Chain for Circular Economy System: A Review." *Journal of Cleaner Production* 108: 409–21. https://doi.org/10.1016/j.jclepro.2015.06.124.

Rajaeifar, Mohammad Ali, Hossein Ghanavati, Behrouz B. Dashti, Reinout Heijungs, Mortaza Aghbashlo, and Meisam Tabatabaei. 2017. "Electricity Generation and GHG Emission Reduction Potentials through Different Municipal Solid Waste Management Technologies: A Comparative Review." *Renewable and Sustainable Energy Reviews* 79 (May): 414–39. https://doi.org/10.1016/j.rser.2017.04.109.

Ripa, M., G. Fiorentino, V. Vacca, and S. Ulgiati. 2017. "The Relevance of Site-Specific Data in Life Cycle Assessment (LCA). The Case of the Municipal Solid Waste Management in the Metropolitan City of Naples (Italy)." *Journal of Cleaner Production* 142: 445–60. https://doi.org/10.1016/j.jclepro.2016.09.149.

Rodić, Ljiljana, and David C. Wilson. 2017. "Resolving Governance Issues to Achieve Priority Sustainable Development Goals Related to Solid Waste Management in Developing Countries." *Sustainability (Switzerland)* 9 (3). https://doi.org/10.3390/su9030404.

Sadef, Y., A. S. Nizami, S. A. Batool, M. N. Chaudary, O. K. M. Ouda, Z. Z. Asam, K. Habib, M. Rehan, and A. Demirbas. 2016. "Waste-to-Energy and Recycling Value for Developing Integrated Solid Waste Management Plan in Lahore." *Energy Sources, Part B: Economics, Planning and Policy* 11 (7): 569–79. https://doi.org/10.1080/15567249.2015.1052595.

Santos, Gilberto, José Salvador, Motta Reis, Fernanda De Oliveira, Silva Maximilian, Espuny Larah, Giffoni Lescura, et al. 2020. "The rapid escalation of publications on covid-19: a snapshot of trends in the early months to overcome the pandemic and to improve life quality." *International Journal for Quality Research* 14 (3): 951–68. https://doi.org/10.24874/IJQR14.03-19.

Sukholthaman, Pitchayanin, and Alice Sharp. 2016. "A System Dynamics Model to Evaluate Effects of Source Separation of Municipal Solid Waste Management: A Case of Bangkok, Thailand." *Waste Management* 52: 50–61. https://doi.org/10.1016/j.wasman.2016.03.026.

Tan, Sie Ting, Haslenda Hashim, Jeng Shiun Lim, Wai Shin Ho, Chew Tin Lee, and Jinyue Yan. 2014. "Energy and Emissions Benefits of Renewable Energy Derived from Municipal Solid Waste: Analysis of a Low Carbon Scenario in Malaysia." *Applied Energy* 136: 797–804. https://doi.org/10.1016/j.apenergy.2014.06.003.

Tan, Sie Ting, Wai Shin Ho, Haslenda Hashim, Chew Tin Lee, Mohd Rozainee Taib, and Chin Siong Ho. 2015. "Energy, Economic and Environmental (3E) Analysis of Waste-to-Energy (WTE) Strategies for Municipal Solid Waste (MSW) Management in Malaysia." *Energy Conversion and Management* 102: 111–20. https://doi.org/10.1016/j.enconman.2015.02.010.

Tozlu, Alperen, Emrah Özahi, and Ayşegül Abuşoğlu. 2016. "Waste to Energy Technologies for Municipal Solid Waste Management in Gaziantep." *Renewable and Sustainable Energy Reviews* 54: 809–15. https://doi.org/10.1016/j.rser.2015.10.097.

Turner, David A., Ian D. Williams, and Simon Kemp. 2016. "Combined Material Flow Analysis and Life Cycle Assessment as a Support Tool for Solid Waste Management Decision Making." *Journal of Cleaner Production* 129: 234–48. https://doi.org/10.1016/j.jclepro.2016.04.077.

Vitorino de Souza Melaré, Angelina, Sahudy Montenegro González, Katti Faceli, and Vitor Casadei. 2017. "Technologies and Decision Support Systems to Aid Solid-Waste Management: A Systematic Review." *Waste Management* 59: 567–84. https://doi.org/10.1016/j.wasman.2016.10.045.

Vučijak, Branko, Sanda Midžić Kurtagić, and Irem Silajdžić. 2016. "Multicriteria Decision Making in Selecting Best Solid Waste Management Scenario: A Municipal Case Study from Bosnia and Herzegovina." *Journal of Cleaner Production* 130: 166–74. https://doi.org/10.1016/j.jclepro.2015.11.030.

Zhou, Ming Hui, Shui Long Shen, Ye Shuang Xu, and An Nan Zhou. 2019. "New Policy and Implementation of Municipal Solid Waste Classification in Shanghai, China." *International Journal of Environmental Research and Public Health* 16 (17). https://doi.org/10.3390/ijerph16173099.

Zorpas, Antonis A. 2020. "Strategy Development in the Framework of Waste Management." *Science of the Total Environment* 716: 137088. https://doi.org/10.1016/j.scitotenv.2020.137088.

INDEX

C

D

E

F

G

H

P

R

S

T

U

V

W

Z